Advances in Pluripotent Stem Cells
http://dx.doi.org/10.5772/intechopen.107777
Edited by Leisheng Zhang

Contributors
Irina Neganova, Vitaly Gursky, Olga Krasnova, Anastasia Kovaleva, Julia Sopova, Karina Kulakova, Olga Tikhonova, Leisheng Zhang, Hao Yu, Xiaonan Yang, Shuang Chen, Xianghong Xu, Zhihai Han, Hui Cai, Zheng Guan, Igor M. Samokhvalov, Anna Liakhovitskaia, Yu-Yun Xiong, Yun-Wen Zheng, Xiuzhi Susan Sun, Quan Li, Guangyan Qi, Paul Disse, Guiscard Seebohm, Nadine Ritter, Nathalie Strutz-Seebohm, Hossein Azizi, Danial Hashemi Karoii

First published in London, United Kingdom, 2024 by IntechOpen
IntechOpen is the global imprint of INTECHOPEN LIMITED, registered in England and Wales, registration number: 11086078, 5 Princes Gate Court, London, SW7 2QJ, United Kingdom
Printed in Croatia

British Library Cataloguing-in-Publication Data
A catalogue record for this book is available from the British Library

Additional hard and PDF copies can be obtained from orders@intechopen.com

Advances in Pluripotent Stem Cells
Edited by Leisheng Zhang
p. cm.

This title is part of the Biochemistry Book Series, Volume 48
Topic: Cell and Molecular Biology
Series Editor: Miroslav Blumenberg
Topic Editor: Rosa María Martínez-Espinosa

Print ISBN 978-1-83769-261-3
Online ISBN 978-1-83769-262-0
eBook (PDF) ISBN 978-1-83769-263-7
ISSN 2632-0983

For EU product safety concerns:
IN TECH d.o.o., Prolaz Marije Krucifikse Kozulić 3, 51000 Rijeka, Croatia,
info@intechopen.com or visit our website at intechopen.com.

IntechOpen Book Series
Biochemistry
Volume 48

Aims and Scope of the Series

Biochemistry, the study of chemical transformations occurring within living organisms, impacts all of the life sciences, from molecular crystallography and genetics, to ecology, medicine and population biology. Biochemistry studies macromolecules - proteins, nucleic acids, carbohydrates and lipids –their building blocks, structures, functions and interactions. Much of biochemistry is devoted to enzymes, proteins that catalyze chemical reactions, enzyme structures, mechanisms of action and their roles within cells. Biochemistry also studies small signaling molecules, coenzymes, inhibitors, vitamins and hormones, which play roles in the life process. Biochemical experimentation, besides coopting the methods of classical chemistry, e.g., chromatography, adopted new techniques, e.g., X-ray diffraction, electron microscopy, NMR, radioisotopes, and developed sophisticated microbial genetic tools, e.g., auxotroph mutants and their revertants, fermentation, etc. More recently, biochemistry embraced the 'big data' omics systems. Initial biochemical studies have been exclusively analytic: dissecting, purifying and examining individual components of a biological system; in exemplary words of Efraim Racker, (1913 –1991) "Don't waste clean thinking on dirty enzymes." Today, however, biochemistry is becoming more agglomerative and comprehensive, setting out to integrate and describe fully a particular biological system. The 'big data' metabolomics can define the complement of small molecules, e.g., in a soil or biofilm sample; proteomics can distinguish all the proteins comprising e.g., serum; metagenomics can identify all the genes in a complex environment e.g., the bovine rumen.

This Biochemistry Series will address both the current research on biomolecules, and the emerging trends with great promise.

Meet the Series Editor

Miroslav Blumenberg, Ph.D., was born in Subotica and received his BSc in Belgrade, Yugoslavia. He completed his Ph.D. at MIT in Organic Chemistry; he followed up his Ph.D. with two postdoctoral study periods at Stanford University. Since 1983, he has been a faculty member of the RO Perelman Department of Dermatology, NYU School of Medicine, where he is codirector of a training grant in cutaneous biology. Dr. Blumenberg's research is focused on the epidermis, expression of keratin genes, transcription profiling, keratinocyte differentiation, inflammatory diseases and cancers, and most recently the effects of the microbiome on the skin. He has published more than 100 peer-reviewed research articles and graduated numerous Ph.D. and postdoctoral students.

Meet the Volume Editor

Professor Leisheng Zhang is a cell biologist and the leader of a research team devoted to tissue engineering and regenerative medicine. In 2012, he obtained a bachelor's degree in veterinary medicine (a BVetMed degree) from Jilin University, China. In 2018, he obtained a Ph.D. in Cellular Biology from the Chinese Academy of Medical Sciences & Peking Union Medical College (CAMS & PUMC) and Tsinghua University, China. In 2021 and 2023, Dr. Zhang completed post-doctoral research at the School of Medicine, Nankai University, China and the Chinese Academy of Sciences (CAS), respectively. In 2024, Dr. Zhang was the central laboratory director and an academic leader at The Fourth People's Hospital of Jinan and the teaching hospital of Shandong First Medical University. He was also a distinguished professor at Gansu Provincial Hospital, and a principal investigator in Shandong Public Health Clinical Center, Shandong University. For decades, Dr. Zhang has undertaken projects funded by the National Natural Science Foundation of China. He has published more than eighty articles and twenty books and has thirty patents to his credit.

Contents

Preface

Human pluripotent stem cells (hPSCs), including human embryonic stem cells (hESCs) and human induced pluripotent stem cells (hiPSCs), are stem cells with self-renewal and multilineage differentiation potential. For decades, we and other investigators in the field have been devoted to verifying the balance of pluripotency and early germ layer specialization of the mesoderm, endoderm, and ectoderm. State-of-the-art literature has put forward the feasibility of hPSCs for the ex vivo, large-scale preparation of stem cells and concomitant functional cells, including hematopoietic stem cells (HSCs), neural stem cells (NSCs), mesenchymal stem/stromal cells (MSCs), and immune cells (e.g., natural killer cells, T cells, megakaryocytes). Meanwhile, numerous studies have indicated the involvement of hPSCs in disease remodeling and drug screening based on diverse preclinical and clinical practices.

In this book, we mainly focus on the definition and the latest updates of hPSCs in regenerative medicine and cell preparation. For instance, MSCs derived from hPSCs reveal preferable bidirectional immunomodulation and long-term in vitro expansion over diverse counterparts from adult tissues such as adipose tissue (AD-MSCs), bone marrow (BM-MSCs), and dental pulp (DPSCs). Overall, the book highlights the feasibility of hPSCs-based cytotherapy for potential clinical applications in the future, including further development of disease remodeling and drug screening.

Leisheng Zhang
The Fourth People's Hospital of Jinan,
Jinan, China

The Teaching Hospital of Shandong First Medical University,
Shandong Province, China

Gansu Provincial Hospital,
Lanzhou, China

Section 1

The Overview of Human Pluripotent Stem Cells

Chapter 1

How Morphology of the Human Pluripotent Stem Cells Determines the Selection of the Best Clone

Vitaly Gursky, Olga Krasnova, Julia Sopova, Anastasia Kovaleva, Karina Kulakova, Olga Tikhonova and Irina Neganova

Abstract

The application of patient-specific human induced pluripotent stem cells (hiPSCs) has a great perspective for the development of personalized medicine. More than 10 hiPSCs clones can be obtained from one patient but not all of them are able to undergo directed differentiation with the same efficiency. Beside, some clones are even refractory to certain directions of differentiation. Therefore, the selection of the "best" or "true" hiPSC clone is very important, but this remains a challenge. Currently, this selection is based mostly on the clone's morphological characteristics. Earlier, using methods of mathematical analysis and deep machine learning, we showed the fundamental possibility for selecting the best clone with about 89% accuracy based on only two to three morphological features. In this chapter, we will expand on how the morphological characteristics of various hiPSCs clones, the so-called "morphological portrait," are reflected by their proteome. By reviewing previously published data and providing the new results, we will highlight which cytoskeletal proteins are responsible for the establishment of the "good" morphological phenotype. Finally, we will suggest further directions in this research area.

Keywords: hiPSCs, hESCs, machine learning, best clone, morphological phenotype, proteome, cytoskeleton

1. Introduction

High-quality clones of human pluripotent cells (hPSCs) are of great importance for research in both basic and translational medicine due to their capacity to differentiate to all cell types of the human body and unlimited self-renewal. Unfortunately, currently available reprogramming methods to generate human induced pluripotent stem cells (hiPSCs) are stochastic, and that causes the presence of a large percentage of partially reprogrammed cells and cells with a low level of pluripotency [1]. The purification of culture is an important requirement to obtain high-quality clones. Usually, this includes either gene expression profiling or evaluation of the cellular morphology by visual inspection or image analysis. However, both of these approaches have limitations. Namely, gene expression profiling gives a direct readout of stemness and differentiation, but it is destructive to cells. At the same time, visual morphological analysis of cells or their images is a nondestructive method, but it is prone to errors and misinterpretation. This explains the urgent need for the development of the noninvasive evaluation of the pluripotent cell cultures, that is. able to link cell morphology to the level of pluripotency. In the first part of the present chapter, we will discuss the morphological features of hPSCs and methods for their automated evaluation.

Currently, there are publications in the literature on the employment of live-cell imaging analysis along with deep machine learning for the development of automated software for the recognition of the best clones. Certainly, these scientific works, discussed in the first part of the chapter, represent only the first attempts of computer image analysis application for the selection of the best clones or identification of cells that have not undergone complete reprogramming. One question remains to be resolved in this method: to what extend can we make general conclusions based on the data from few studies, even with a large number of samples?

More than 500 distinct human embryonic stem cell lines (hESCs) have been generated to date, but only less than 100 lines are available now for general research as fully characterized lines (NIH stem cell registry, https://grants.nih.gov/stem_cells/registry/current.htm). In addition, there are multiple patient-specific human induced pluripotent stem cells (hiPSCs) lines and the list of these lines continues to grow. By 2020, about 131 studies were classified as clinical trials involving human pluripotent stem cells (hPSCs, comprising both hiPSCs and hESCs) [2]. The analysis published by Deinsberger and colleagues [2] revealed that the number of clinical trials involving hiPSCs was substantially higher than the one involving hESCs (74.8% vs. 25.2%). However, when counting only interventional studies, it appears that the majority (73.3%) was done with the use of hESCs. Application of patient-specific hiPSCs helps to overcome both ethical and immunological issues but hESCs are still widely used in the field of translational and regenerative medicine, disease modeling, and drug screening. Importantly, both hPSC types are very similar in their morphological characteristics not being molecular equivalents [3, 4].

Regardless of the common morphological features of hiPSCs and hESCs, it is well-documented that major line-to-line morphological variability exists even in the same culture conditions and with the use of the same propagation technique [5]. This fact raises the question of whether there are common morphological features that distinguish a "good" hPSCs clone from a "bad" one. Finding the answer to this question is extremely important as the maintenance of hPSCs in culture is not only expensive but it is also very labor intensive. The development of an automated quality control protocol can improve the utility of the high-quality clinical-grade cells.

A noninvasive method of visual inspection of the morphological appearance remains the main criteria used to select the best hPSCs clone. However, until now, it was not clear which parameters of morphology are closely associated with the pluripotent state.

Recently, we analyzed morphological parameters of several hPSCs lines of various passages. We first extracted the parameters from phase-contrast images and constructed classification models of colonies by morphological phenotype [6], and then we used image analysis with convolutional neural networks (CNNs) [7]. Further to this, expression analysis of 11 pluripotency markers genes allowed us to identify phenotype-specific sets of genes that could be used for the selection of the best clones, meaning the fundamental possibility of constructing a morphological "portrait" of a colony informative for its automatic identification. Additionally, we performed a proteomic analysis of several hPSCs samples from various lines used before for the computational analysis and showed that cells with different phenotypes from various lines cluster at the proteome level in accordance with their morphological phenotype [7].

Multiple studies have provided datasets of comparative proteomes of various hPSC lines. Several studies used proteomic approaches to find proteins that regulate pluripotency [8–10] or to conduct a comparative proteomic analysis of supportive and unsupportive matrix substrates for hESCs maintenance [11]. In addition, several papers described quantitative proteomic analysis of hESCs differentiation [12, 13]. However, only in 2019, the first paper appeared on the analysis and comparison of the proteomic landscapes of 20 hiPSCs lines classified as stable and unstable based on colony morphology. This study has shown that different morphological "portraits" of colonies are associated with different proteomic profiles and different competencies for directed differentiation [14]. Furthermore, it has been shown that a direct relationship exists between pluripotent markers (DNMT3B, DPPA4, SALL4, CD9) and morphological "portraits" of various lineages [6, 14].

In this chapter, we will review the current knowledge about how automated evaluation of the morphological portrait is used to control the hPSC phenotype, and how it is connected to the proteomic analysis. Next, we will present our own proteomic data analysis of hPSCs in respect to their morphological phenotype. We will pay special attention to the cytoskeleton proteins, as some of them turned out to be the top candidates in determining the best cell and colony morphology. The future will tell us if the hiPSC technology will ultimately overcome the current challenges and will finally make its way into routine clinical application with the help of automated recognition of the best clone based on the morphological selection.

2. Morphological features of human pluripotent stem cells and methods for their automated evaluation

Currently, work with hPSCs begins with the assessment of their morphology by an expert to determine if there are signs of spontaneous differentiation or other unwanted changes. Established standard criteria for morphological features of hPSCs during their expansion can be described as: (a) a high nucleus/cytoplasm ratio, (b) prominent nucleoli, (c) formation of compact and round colonies with flat and densely packed cells with scant cytoplasm. Additional important marker is the presence of a clear and smooth colony edge [6, 15, 16]. As hPSC colonies propagate in culture, cells might spontaneously deviate from pluripotency toward a differentiated state. In that case, cell morphology changes dramatically, and it is very noticeable; the cells in the colony start to distribute sparser, the distance between the cells expands and cells significantly increase in size, undergoing a characteristic shape change [6]. In addition, undifferentiated hPSCs have more relaxed chromatin than differentiated; during the differentiation process, nucleoli become unclear and invisible under phase contrast microscopy [16]. Notably, only a very skilled expert can notice this alteration; therefore, evaluation of the cultures by the observation of the colonies morphology by an expert obviously depends on the expert's skills. Undoubtedly, the safe application of hPSCs in the clinic requires the creation of a cell evaluation method, which would be less dependent of the expert's skills.

In recent years, several image analysis approaches have been developed. Machine learning, which involves pattern recognition and computational learning, is one of the most widely used strategies. In addition to pattern recognition, some of machine-learning algorithms classify cells into several quality classes, which are related to non-morphological image features, such as the distribution of luminance intensity. The fully automated system has been reported for morphology-based evaluation of iPSC cultures that consists of time-lapse microscopy and image analysis software [17, 18]. The system acquires low-light phase-contrast images of iPSC growth collected during a period of several days, measures geometrical- and texture-based features of the colonies throughout time, and derives a set of six biologically relevant features to automatically rank the quality of the cell culture. This method has shown that hiPSCs that are classified visually could be adequately distinguished with local binary patterns and an intensity histogram [18]. The classifier presented in that work successfully identifies different cell stages for a wide range of scenes that can include different-sized colonies, varying amounts of dead cells and debris, and differentiated cells within colonies [18].

As mentioned before, in case of cell differentiation, nuclear structures reconfigure dynamically. The method published by Tokunaga and colleagues [19] for discrimination of the bona fide hiPSCs from non-reprogrammed ones, is based only on the fine differences of the nuclear morphology between cells. Namely, this work has demonstrated that specific quantitative parameters contributing to morphological discrepancies reside in the nuclear sub-domains. Analysis of nuclear morphologies revealed dynamic and characteristic signatures, including the linear form of the promyelocytic leukemia (PML)-defined structure in hiPSCs, which was reversed to a regular sphere upon differentiation. Thus, this data confirmed that hiPSCs have a markedly different overall nuclear architecture that may contribute to highly accurate discrimination based on the cell reprogramming status [19].

Similarly, the paper by Kato et al. [20] demonstrated a noninvasive image-based evaluation method for detecting partially differentiated colony morphology in heterogeneous colony populations *via* live image analysis. The authors analyzed eight major parameters comprising 27 sub-parameters selected as essential for further analysis of 303 hiPSC colonies. The data showed a relationship between image features and gene expression by analyzing the expression of hiPSC colonies classified by using spatial frequency. Next, colony morphology classification based on the statistical analysis of the live-cell images with unbiased morphological parameters was compared with classification based on global gene expression profiles of individual colonies. Classification utilizing statistical analysis produced similar results as compared to classification based on gene expression profiles. Thus, authors concluded that quantitative morphological evaluation method facilitates the noninvasive analysis of hiPSC conditions and demonstrates its utility in recognition hPSCs heterogeneity.

The paper by Wakui and colleagues [17] aimed at establishing quality classification of hiPSC images into three classes (poor, moderate, and good) by evaluating the biological features used in the visual inspection. Three features associated with biological structures such as the number of nucleoli, the crack area rate, and the differentiating cellular nuclei area rate were chosen by the expert. Importantly, these features were effective for quality evaluation by the visual inspection. As mentioned before, the number of nucleoli is a feature indicating a non-differentiated state, and cells with many nucleoli are considered to be of good quality. In contrast, the crack area rate and the differentiating cellular nuclei area rate are indicators of deviation from pluripotency. The method identifies three feature detectors and the cell quality classifier, the inputs of which are the outputs of the detectors. Then, in the image analysis method, the feature detectors and the classifier are applied to each of the regions of interest (150 pixels, 50 μm) of a phase contrast image. For machine learning of the nucleolus detectors, the nucleoli dataset was used as training data. The crack detector and the differentiating cellular nuclei detector were tuned with the masked dataset. The cell quality classifier was developed with the labeled dataset. Nucleoli observed in undifferentiated cells are nearly oval-shaped, 3- to 6-μm in diameter, and appeared black under phase-contrast observation. For confirming the classification capability of these three features, the distributions of the features for each cell quality class of a respective cell line were investigated and the accuracy for cell quality classification that was equivalent to visual inspection with respect to the three hiPSC lines was confirmed [17].

Interestingly, the paper by Nishimura and colleagues remains the only paper, which is based upon the morphology of a cellular organelle; it describes the use of the mitochondria distribution and state for distinguishing reprogrammed mouse PSCs [21]. The authors reported the development of an imaging system, termed phase distribution (PD) imaging system, which visualizes subcellular structures quantitatively in unstained and unlabeled cells. The PD image and its derived PD index reflected the mitochondrial content, enabling quantitative evaluation of the degrees of somatic cell reprogramming and mouse PSCs differentiation [21]. The dynamic changes in mitochondrial biogenesis and antioxidant enzymes are well-documented during the spontaneous differentiation of hESCs, as well as during the reprogramming process [22]. Unlike in PSCs, in the somatic cells mitochondria are numerous and large, reflecting their dependence on oxidative phosphorylation for efficient energy production. It is well-known that the reprogramming of the somatic cells into iPSCs is accompanied by a metabolic shift from oxidative phosphorylation to glycolysis, concomitant with changes in structure and function of mitochondria [23, 24]. Indeed, iPSCs that are reprogrammed to different degrees show an inverse relationship between their pluripotency and mitochondrial activity [25]. Thus, morphological changes of subcellular structures such as mitochondria may serve as an additional useful marker to evaluate the pluripotency of reprogrammed cells.

Our own data on morphological parameters from three lines (hESC line H9, hiPSC line AD3, and hiPSC line HPCASRi002-A) revealed that several morphological criteria can be used to distinguish between “good” and “bad” phenotypes (**Figure 1A**) [6], thus demonstrating that these are strong and reliable criteria for determining the phenotype of hPSCs. We tested seven morphological parameters in total as possible predictors in the neural network-based classification models of the colony phenotype. The models aimed to predict the probability of the colony phenotype (either ‘good’ or ‘bad’) and were trained on the morphological parameter values of colonies or cells. A minimal model was selected for each data type that contained a minimal number of predictors and still provided the prediction accuracy close to that in the model with all predictors included. For the colony morphology data, we found a minimal model of four input parameters (Perimeter, Minor Axis, Shape Factor, and AIS) that showed 74% accuracy on average, while only two parameters (Perimeter and Shape Factor) were enough to provide a 68% average accuracy in the minimal classification model for the cellular morphological data (**Figure 1A**).

As an alternative approach for the colony phenotype prediction, we applied convolutional neural networks (CNNs) directly to the phase-contrast images of colonies, omitting the intermediate step of extracting the morphological parameters from the

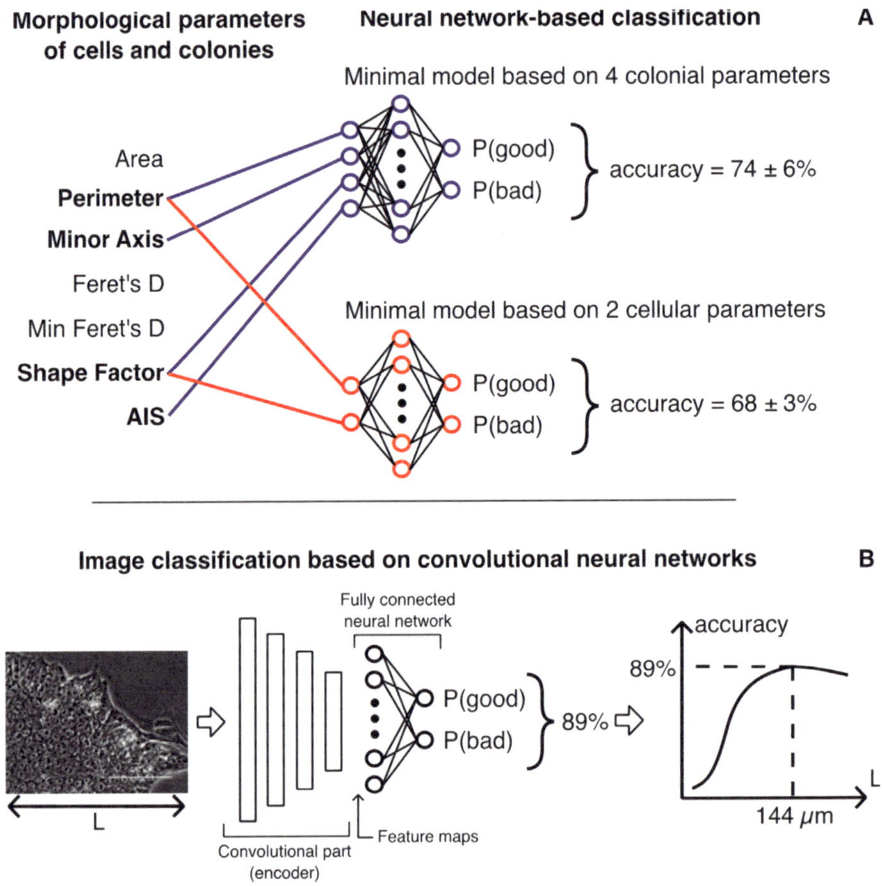

Figure 1.
Two approaches to colony phenotype prediction using automatic classification [6, 7]. A: Minimal classification models based on neural networks use four morphological parameters of colonies or two morphological parameters of cells as predictors. The corresponding parameters are shown in bold and connected with the classifier input with blue or red lines for the colonial or cellular data-based models, respectively. The output of the models is the probabilities that the colony will have a good or bad phenotype (P(good) or P(bad), respectively). B: CNN trained directly on colony images. The convolutional part of the CNN extracts the most representative feature maps from the images. These features are then passed to the input of a fully connected network trained to separate the phenotypes. CNNs trained on datasets with images of various linear size L show the prediction accuracy that has a maximum at the size L ~ 144 μm, which can be interpreted as the most informative spatial scale. AIS, area of intercellular space (for colonies only).

images (**Figure 1B**). CNNs extract the informative features, called "feature maps," from the images and use these features to predict the phenotype at the output. We trained the CNN-based classification model on phase-contrast images of the H9 hESC line colonies and obtained an 89% accuracy in phenotype prediction [7].

The morphological "portrait" of the colony that can be associated with clonality is a complex trait, with no clear spatial scale that could unambiguously separate the natural morphological variability within the colony from signs of clonality loss. Trying to answer the practical question about the most informative spatial scale at which the colony phenotype could be recognized by the automated classifiers most effectively, we trained CNNs on multiple datasets containing images of various linear size. We found an optimal image size of ~144 μm providing the best classification

Morphological data	Phenotypes	Classification data	Performance	Other details	Source
Six morphological parameters extracted from the analysis of hiPSC colonies generated from human fibroblasts using two different reprogramming strategies	12 colony types based on morphological and gene expression patterns, across which non-ESC-like colonies were contrasted against iPSC colonies	-	Criteria to determine iPSC colonies were formulated in terms of specific values (mean ± s.d.) of morphological parameters and demand for cells to be positive for endogenous Sox2 and Cdx2	First evidence that the expression of either Oct3/4 or Nanog is not appropriate for the identification of iPSCs, but rather the expression of endogenous Sox2 and Cdx2 is a reliable marker of iPSCs	Wakao et al. [27]
Six geometric- and texture-based features of iPSC colonies extracted from time-lapse videos of 12 clones from four iPSC lines	Good, fair, and poor	Classification tree based on the probability distribution of six features per phenotype value	Accuracy = 0.80–0.89	The degree of cell compaction and the doubling time were shown to be the features with the highest predictive power	Maddah et al. [18]
Automatically extracted features from phase contrast images for iPSC lines (201B7 and 253G1), newly generated iPSC lines (1H–4H), non-iPSC lines (15B2 and 2B7), and somatic cells (human mammary epithelial cells, HMECs)	Properly and improperly reprogrammed cells (iPSCs vs. non-iPSCs)	Supervised machine learning algorithm wndchrm (weighted neighbor distances using a compound hierarchy of algorithms representing morphology)	Accuracy = 0.87–0.96	Specific morphology quantification suggested that signals contributing to morphological discrepancies reside in nuclear sub-domains	Tokunaga et al. [19]
128-dimensional vector of features extracted by the scaled invariant feature transformation (SIFT) applied to phase contrast images of hiPSCs.	Good, semigood, and bad	*k*-nearest neighbors, multiclass support vector machines, and other classification methods	Accuracy up to 0.62	*k*-nearest neighbors method showed the best performance	Joutsijoki et al. [28]

Morphological data	Phenotypes	Classification data	Performance	Other details	Source
120 colony morphology parameters (reduced to 27 parameters) extracted from live-cell phase contrast images of aberrant 201B7-1A subclone (with unusual undifferentiated ESC- like colony morphology), its healthy parent 201B7 line (with typical ESC-like colony morphology), 253G1 cell line and its subclone 253G1-B1	No phenotyping. Colonies were clustered into five major clusters ('morphological categories') according to their morphological parameters	Unsupervised hierarchical clusterization of colony morphological parameters	-	Cluster analysis results for colony morphologies were reproduced by individual gene expression profiles	Kato et al. [20]
3 colony features (number of nucleoli, the crack area rate, and the differentiating cellular nuclei area rate) extracted from phase contrast images of MRC5, Edom, and 201B7 hiPSC cell lines	Good, moderate, and poor	Supporting vector machine	Accuracy = 0.86	Image analysis framework was created for automated extraction of the three features and classifier application to regions of interest for a phase contrast image	Wakui et al. [17]
Full data comprise low-magnification phase-contrast images and a fluorescence channel for alkaline phosphatase staining of mouse ESC colonies. The only morphological parameter used for phenotype classification is colony circularity.	Pluripotent colonies, mixed colonies, and differentiated cells	Pluri-IQ software, which can automatically quantify the percentage of pluripotent, mixed, or differentiated cells through culture images, with the following cascade modules: segmentation, machine learning (random forest classifier), validation, and automatic scoring	Accuracy > 0.90	Pluri-IQ uses as input large images and has advantages compared with other similar software	Perestrelo et al. [29]

Morphological data	Phenotypes	Classification data	Performance	Other details	Source
Authors derived a phase distribution (PD) index and a 3D extracted PD (ePD) image from an improved retardation modulated differential interference contrast (RM-DIC) imaging systems, applied to mouse iPSCs. Both the PD index and ePD image were designed to reflect subcellular structures, specifically the mitochondrial content.	Qualitative phenotypic states of different degrees of somatic cell reprogramming, ESC differentiation, and pluripotency	-	Various quantitative measures for correspondence between the PD index or ePD image with various phenotypic values	The PD index and ePD image were shown to reflect different degrees of somatic cell reprogramming, ESC differentiation, and pluripotency	Nishimura et al. [21]
Automatic feature maps extraction as a part of classification model training for phase contrast images of diseased iPSCs expressing the Huntington's disease phenotype	Four morphological classes: debris, dense, differentiated, and spread	CNN, with data augmentation by supplementing a minimal biological dataset *via* image generation using generative adversarial networks. Model based on Markov chain stochastic processes is used to account the influence of temporally constrained differentiation on classification model training.	Recall (true positive rate) = 0.88–0.94	Presented work highlights the importance of exploiting temporal relationships between image classes, which is an example of using 'domain knowledge' in combination with deep learning	Witmer and Bhanu [30]
Seven morphological parameters of colonies and cells from phase-contrast images for H9 hESC line, AD3 hiPSC line, and HPCASRi002-A hiPSC line	Good and bad	Neural networks	Accuracy = 0.74 ± 0.06 for colonial parameters and 0.68 ± 0.03 cellular parameters	Four parameters for colony morphology and two parameters for cellular morphology are enough	Krasnova et al. [6]

Morphological data	Phenotypes	Classification data	Performance	Other details	Source
Feature maps extracted from phase-contrast images of iPSC clones derived from peripheral blood mononuclear cells	Undifferentiation, cracked, built-up, differentiation	Feature maps extraction by VQ-VAE encoder (unsupervised learning) + support vector machine for phenotype classification based on extracted feature maps	Accuracy = 0.89	Support vector regression method was used to predict the expression of 218 genes based on feature maps extracted from images, with $0.3 < R^2 < 0.69$	Wakui et al. [31]
Feature maps are automatically extracted from phase contrast images for H9 hESC line during model training	Good and bad	CNN	Accuracy = 0.89	Most informative spatial scale was determined	Mamaeva et al. [7]

Table 1.
Selected results of phenotype classification based on morphological features of hPSCs colonies.

performance (**Figure 1B**). This size is intermediate between typical colony and cell sizes, reflecting the fact that both cellular and colonial information should be taken into account.

We linked these results to the transcription studies by measuring the expression of 10 pluripotency markers (*DNMTB3, SALL4, IGFR1, CD9, DPPA4, OCT4, REX1, NANOG, SOX2,* and *KLF4* genes) in colonies from the hESCs (H9) and hiPSC (AD3 and HPCASRi002-A) lines with different phenotypes [6]. We found that the *SALL4, DNMTB3, REX1, DPPA4*, and *SOX2* genes demonstrated differentiated expression in colonies of different phenotypes, thus confirming that the phenotypes did represent the pluripotency status of the colony.

Finally, the question can be asked, whether the cultivation conditions, namely various culture media and matrixes, affect the morphological parameters important for morphological phenotype recognition. By other worlds, is evaluation method based on the morphologies of various hPSC lines applicable under different culture conditions? The paper by Harkness and colleagues [26] addresses the effect of five different media, namely, mTESR1, Essential E8, StemPro (SP), mouse embryonic fibroblasts conditional media (CM) and StemMacs iPS-Brew XF (SM), on the morphological parameters of the three established hESCs lines (MEL1, WA09, ESI-hES3). These lines were routinely grown on Matrigel (Corning) in mTESR1 media before switching to a different media. As a result, the authors observed distinct and measurable differences in nuclear and cell morphology between different culture conditions. In CM and SP cultures, authors noticed a looser colony structure and a flatter appearance when compared to mTESR1, E8, or SM media. The morphological parameters such as nuclear area, cell area, cell roundness, and cell spread in all three lines demonstrated an overall decrease, while in the least defined media, CM, in all cell lines the cells became larger. Moreover, the nuclear/cytoplasmic ratio varied between the lines, suggesting that media composition can affect the cell's parameters and may cause cytoskeletal remodeling. Furthermore, high content imaging demonstrated that hESCs grown in different media exhibit significantly different cytoskeletal architecture while maintaining their pluripotent status, suggesting that cytoskeleton has become more stable in xeno-free media [26]. Thus, the detailed analysis provided in this research let to conclude that morphological alterations of cell phenotype can be associated with the changes of cell culture conditions. However, it can be assumed that when changing from culture medium to another, cells undergo a period of adaptation and, perhaps, after a certain number of passages, they will restore their previous morphological parameters. However, this needs further verification.

Thereby, to create a reliable system for recognizing the best clones, further studies of different hPSC lines during their cultivation on various matrixes and media are required. The creation of a single database that combines data on morphological parameters from numerous lines will improve the methods of automatic clone's recognition for their reliable application in clinic.

In **Table 1**, we summarized some findings about phenotype classification based on morphological features of hPSCs colonies.

3. Morphological phenotypes of the hPSCs reflected in different proteomic landscapes

Proteomics analysis provides an excellent tool for large-scale quantification and benchmarking of cells and an opportunity to understand deeper the rules that govern hPSCs morphology. Compared to other ~omics, such as transcriptomics and genomics approaches, proteomics analysis measures the translated proteins. Most of the previous studies have used proteomics approaches to identify proteins important for stem cell pluripotency maintenance and for lineage differentiation [8–10, 13]. Some studies have explored the membrane proteins [32] or the hESCs phosphoproteome [8, 9] in comparison to hiPSCs proteome and phosphoproteome [33]. In addition, molecular differences of the proteome level between hiPSCs of different somatic origin were described [10]. A comparative proteomic analysis has been published comparing supportive and unsupportive extracellular matrix substrates used for hESCs maintenance [11]. All these studies revealed a huge number of proteins known to be important for hPSCs maintenance, namely, cell cycle and DNA damage repair proteins, proteins involved in integrin binding, intracellular vesicle trafficking proteins, RNA binding, adaptor proteins and histones, proteins of exosomes biogenesis and tumorigenesis, zinc finger proteins, mitochondrial proteins, and many others.

The goal of our study was to analyze the hPSCs proteome in accordance with the selected morphological phenotypes [6, 7]. Thus, we compared a proteomic "portrait" of the "true" or the best hPSC colonies versus the "bad" ones.

In the hESCs (H9) samples, we have identified in total 1791 proteins in a good agreement with the Van Hoof and colleagues [9] who have identified 1775 proteins from undifferentiated hES cell line HES-2. Our data demonstrated a clear separation of the samples in accordance with their morphological phenotypes [7], in agreement with the previously published data of Bjørlykke and colleagues [14], thus emphasizing that good and bad morphological populations are molecularly distinct. Comparative proteome analysis of the hESC (H9) colonies with the good morphological portrait compared to colonies with poor morphology and signs of spontaneous differentiation showed that 63 proteins are downregulated and 25 proteins are upregulated (**Figure 2**) [7].

In the context of the identified proteins that determine the morphology of hPSCs, we were especially interested in cytoskeletal proteins since they form the structural network of the cell. In addition, the migration and spread of motile cells, such as hPSCs, over the surface of the substrate accompanied by the reorganization of their actin network. Among 25 upregulated proteins, four belong to cytoskeletal proteins (MYH7, RDX, CNN3, and AIF1L). The other one is the tight junction protein 1, or ZO1 (TJP1), one of the functions of which is to organize the components of tight intermediate junctions and bind them to the cortical actin cytoskeleton. In our analysis, MYH7 appears on the top position among the upregulated proteins (**Figure 2**). The MYH7 gene, known as myosin beta heavy chain (MHC-β), is classified as a type I fiber. Myosins are a large family of proteins that share the common features of ATP hydrolysis, actin binding, and potential for kinetic energy transduction. They composed of a pair of myosin heavy chains (MYH) and two pairs of nonidentical light chains. At least 10 different MYH isoforms have been described in mammalian cells, and the role for the identified in hESCs MYHs, such as MYH16, MYH15, MYH10, MYH9, and MYH7, is aviating to be discovered. This protein is a critical component of the sarcomere's structure and interacts with other key cytoskeletal proteins such as actin, troponin, and myosin-binding protein C (MYBPC3). Its role was shown in directed differentiation of hiPSCs into cardiomyocytes [34] but has not been studied in hPSCs. The dynamics of actin-myosin contraction are directly regulated by the amount of alpha-actinin-3 (ACTN3), which forms cross-links, and the absence of ACTN3 disrupts the symmetry of the actin network in cells. Human ESCs exhibit basal-apical polarity, junctional complexes, integrin-dependent matrix adhesion, and E-cadherin-dependent adhesion, all of which are characteristics of the epiblast epithelium of a mammalian embryo.

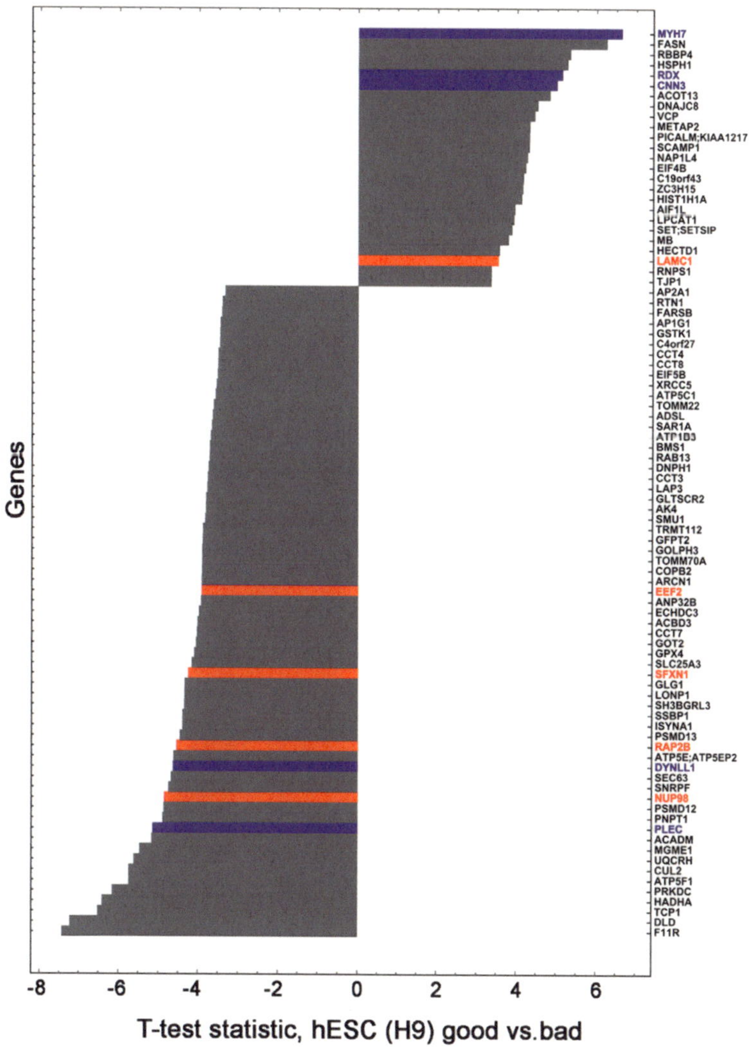

Figure 2.
Z-score-ranked distribution plot for the proteins of the hESC (H9) colonies with the "good" morphological portrait compared to colonies with "bad" morphology. Cytoskeletal proteins are marked blue, and proteins identified via comparison of the "good" morphological hESC H9 samples versus two hiPSC lines with the same characteristics are marked red.

When hPSCs are subject to enzymatic digestion during propagation of the colonies, epithelial structures are destroyed, which leads to programmed cell death; here, actin-myosin contraction is a critical effector of the cell death response to enzymatic dissociation [35]. With this regard, inhibition of the myosin heavy chain ATPase, inhibition of the myosin heavy chain, and inhibition of the myosin light chain (MLC) have been shown to increase the survival and cloning efficiency of individual hPSCs [36]. ROCK inhibition decreases phosphorylation of MLC, suggesting that inhibition of actin-myosin contraction is also the mechanism through which ROCK inhibitors increase cloning efficiency of hESCs [37]. In addition, ROCK1/ROCK2 silencing demonstrated that ROCKs regulate MYH function through MLC phosphorylation in hESCs, which, in turn, leads to membrane blebbing and cell death [36]. Lastly, MYH9 and MYH10 are the most highly expressed MYHs with the conserved sites in hESCs. Treatment of hESCs with MYH9/MYH10 siRNAs demonstrated severe phenotypic changes after 96 hours of transfection but increased cell attachment, survival, and cloning efficiency [36]. On the other side, as mentioned above, MYH7 is regarded as a mesenchymal and specifically myocardial marker gene [38]. Our data is also in a good agreement with the earlier work, which demonstrated that a high level of MYH7 protein was detected in hESCs but not in hiPSCs, while MYH9 was identified in both cell types [12]. Obviously, the role of MYH7 needs to be elucidated further. We were able to detect a very thin network of MYH7 colocalized with F-actin fibers but observed its destruction in hESCs (H9) clones with bad morphology (**Figure 3**); supporting the data from

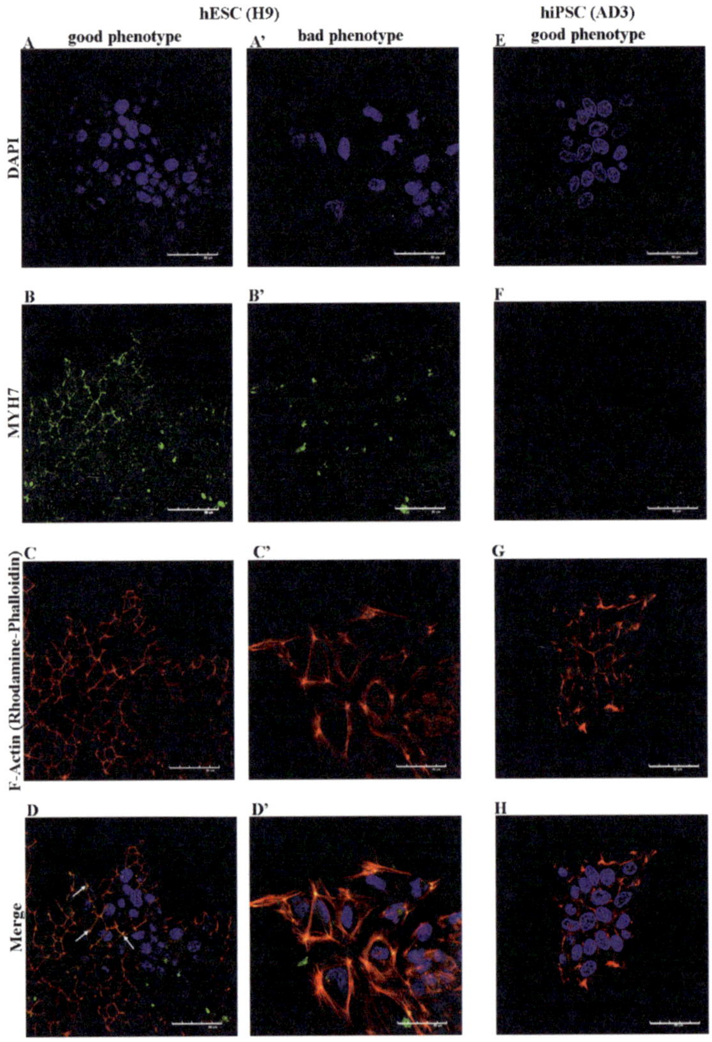

Figure 3.
Colocalization of the MYH7 with F-actin in the organized cytoskeleton network in colonies with "good" phenotype and complete loss of such structural organization and colocalization in "bad" hESCs. Thin white arrows are pointing on the colocalization of F-actin and MYH7. Immunofluorescence was performed using anti-MYH7 antibodies (Santa Cruz Biotechnology, sc-53,089), secondary anti-mouse IgG tagged with Alexa Fluor 488 (Abcam, ab150113), and Rhodamine-Phalloidin (Invitrogen, R415). Scale bar 50 μm.

proteome study, we did not detect MYH7 staining in hiPSCs. The emergence of MYH7 as a top candidate to support the best hESCs morphology might reflect the complexity of the hESCs proteome.

Radixin (RDX) was ranked fourth in our list (**Figure 2**). Radixin is a cytoskeletal protein that may play an important role in binding actin to the plasma membrane. Its exact role for hPSCs has not been explored. However, cellular functions, such as migration and adhesion, require a highly dynamic cytoskeleton behavior. Linker proteins of the ERM family (ezrin/radixin/moesin) can interact with both F-actin and several transmembrane proteins, providing a connection between extracellular cues and the cytoskeleton. The involvement of ERM proteins in a variety of cell functions in the embryonic and early postnatal brain, including axonal outgrowth, morphological rearrangement, cell migration, and signaling, have been described [39]. It is important to note that radixin has been shown to concentrate in the cleavage furrow of dividing cells and may have a role in proliferation [40, 41], the high speed of which is important for pluripotency maintenance [42].

Lastly, throughout the top represented cytoskeletal proteins in hESCs with a good morphological phenotype, we want to discuss CNN3, Calponin 3 (**Figure 2**). Calponin is an actin filament-associated regulatory protein expressed in smooth muscle and multiple types of non-muscle cells. It is capable of binding to actin, calmodulin, and tropomyosin. Three homologous genes, *CNN1*, *CNN2,* and *CNN3*, encoding calponin isoforms 1, 2, and 3, respectively, are present in vertebrates. All three Calponin isoforms are actin-binding proteins with functions in inhibiting actin-activated myosin ATPase and stabilizing the actin cytoskeleton, while each isoform executes different physiological roles based on their cell type-specific expressions. Calponin 1 (*CNN1*) is specifically expressed in smooth muscle cells and plays a role in smooth muscle contractility. Calponin 2 (*CNN2*) is expressed in both smooth muscle and non-muscle cells and regulates multiple actin cytoskeleton-based functions. Calponin 3 (*CNN3*) participates in actin cytoskeleton-based activities in embryonic development and myogenesis. Experiments with cytotrophoblasts from human placenta demonstrated that *CNN3* gene knockdown promoted actin cytoskeletal rearrangement, suggesting *CNN3* to be a negative regulator of trophoblast fusion [43]. With the course of trophoblastic cell differentiation, CNN3 undergoes downregulation. In the trophoblastic cells, membrane flexibility is necessary for membrane fusion [43]. However, whether *CNN3* expression affects the flexibility of the hPSCs plasma membrane is not known, but it may be suggested that regulation of actin cytoskeletal rearrangement by *CNN3* is required for hPSCs. Recently, Calponin 3 was studied in the U2OS osteosarcoma cells, where RNAi knockdown studies revealed that CNN3 is a dynamic component of stress fibers and is required for controlling proper contractility of the stress fiber network [44]. Importantly, the role for CNN3 was also shown for the maintenance of the lens epithelial phenotype where downregulation of CNN3 expression induced changes in cell shape, reorganization of actin cytoskeleton, and formation of focal adhesions resulting in activation of mechanosensitive transcription factor Yap in association with decreased E-cadherin and β-catenin expression [45]. Whether or not the high level of CNN3 in hESCs is associated with the focal adhesion and E-cadherin maintenance remains to be elucidated. Our immunofluorescence study supported obtained proteomic data and revealed a colocalization of the CNN3 with F-actin in the organized cytoskeleton network in colonies with good morphological appearance and complete loss of such structural organization and colocalization in "bad" hPSCs (**Figure 4**).

Among the top downregulated cytoskeletal proteins in hESCs with good morphology appeared DYNLL1 (Dynein light chain 1) and PLEC (PLECTIN) (**Figure 2**).

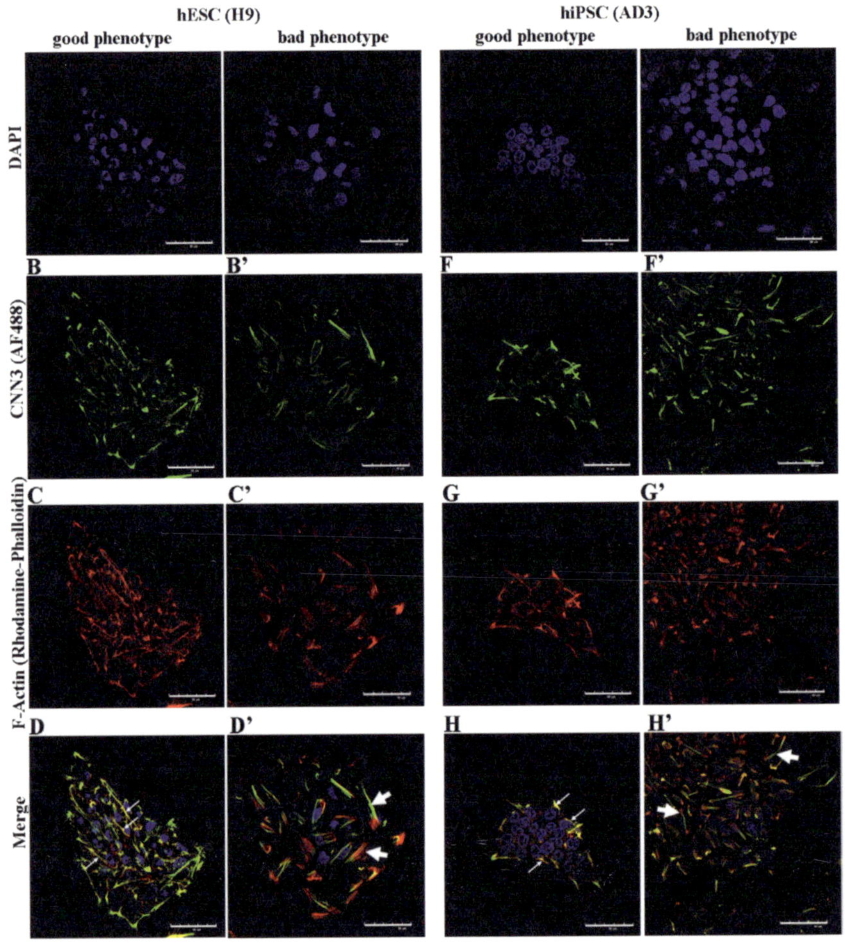

Figure 4.
Colocalization of the CNN3 with F-actin in the organized cytoskeleton network in colonies with "good" phenotype and complete loss of such structural organization and colocalization in "bad" hPSCs. Thin white arrows are pointing on colocalization of F-actin and CNN3; thick white arrows show the absence of colocalization. Immunofluorescence was performed using anti-CNN3 rabbit antibodies (ATLAS, HPA051237), secondary anti-rabbit IgG tagged with Alexa Fluor 488 (Abcam, ab150077), and Rhodamine-Phalloidin (Invitrogen, R415). Scale bar 50 µm.

Cytoplasmic DYNEIN1 acts as an engine for intracellular retrograde mobility of vesicles and organelles along microtubules. Plectin maintains tissue integrity and associate with intermediate filaments (IF). It acts as a cytoskeletal cross-linking agent and signaling scaffold, influencing both the mechanical and dynamic properties of the cytoskeleton. As a member of the cytolinker protein family, plectin has a multidomain structure that is responsible for its ability to bind to many cytoskeletal proteins. It binds not only to all types of IFs, actin filaments, and microtubules but also to transmembrane receptors, nuclear envelope components, and several kinases with known roles in cell migration, proliferation, and energy metabolism. The exact role of plectin in cytoskeletal dynamics is not studied for hPSCs, but in view of its downregulation for a good morphological phenotype, it can be assumed that lower level of protein expression may play a role in the cytoskeletal plasticity of these cells.

In **Figure 5A**, we demonstrate biological processes that we have identified to be related to the upregulated proteins in hESCs H9 line with good morphology. As can be seen, among the most important of them are cellular component biogenesis and assembly, organelle organization, epithelium development, cytoskeleton organization with DNA packaging, and chromatin organization. It is important to highlight that among the most important processes are up-regulation of actin filament-based processes and actin cytoskeleton organization. Among biological processes associated with downregulated proteins (**Figure 5B**), we identified cellular metabolic processes, nitrogen compound metabolic processes, cellular localization, protein transport, DNA metabolic processes, and many others related to control of the cellular metabolism. Also, cellular component analysis (**Figure 5C**) revealed cytoplasm, actin cytoskeleton,

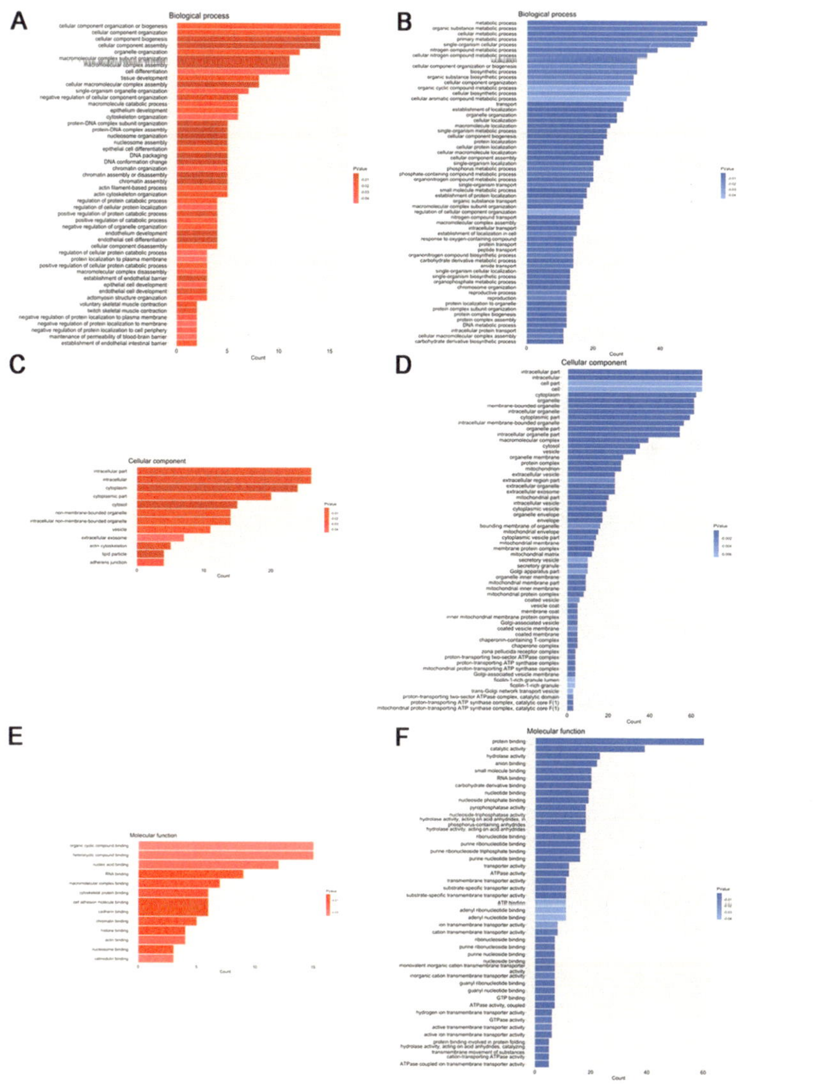

Figure 5.
Biological functions, cellular functions, and molecular processes associated with the up- and downregulated proteins in hESCs H9 cell line. Red bars (A, C, and E) refer to upregulated proteins; blue bars (B, D, and F) refer to downregulated proteins.

and adherence junction among most upregulated processes, while cellular processes related to cell, membrane-bounded organelle, vesicle, mitochondrion, mitochondrial, and organelle envelope appeared to be downregulated (**Figure 5D**). Importantly, molecular functions associated with "good" morphology include cytoskeletal protein binding, cell adhesion molecular binding, cadherin binding, actin binding together with chromatin and histone binding (**Figure 5E**), while among downregulated cellular functions appeared ribonucleotide, purine nucleotide binding, ATP binding, and GTP binding (**Figure 5E**).

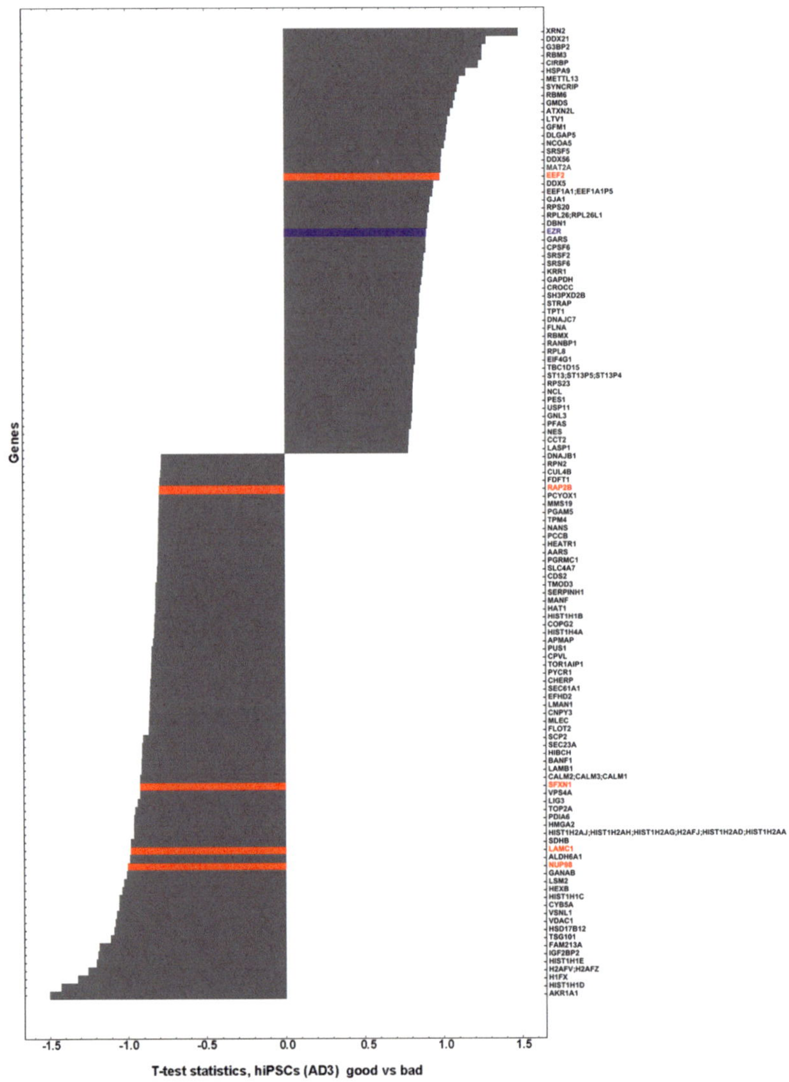

Figure 6.
Z-score-ranked distribution plot for the proteins of the hiPSC (AD3) colonies with the "good" morphological portrait compared to colonies with "bad" morphology. The EZRIN protein is marked blue, and proteins identified via comparison of the "good" morphological hESC H9 samples versus two hiPSC lines with the same characteristics are marked red.

In addition to hESC line H9, we analyzed by the same proteomic approach two hiPSC lines of different origins, namely obtained from the neonatal fibroblasts line AD3 and a patient-specific hiPSCs line HPCASRi002-A (CaSR) [6, 7]. Morphological evaluation of the "good" and "bad" hiPSC clones, as well as comparisons of their proteomic landscapes was performed as for hESC and H9 samples [6, 7]. Interestingly, in the same analysis of proteins associated with cytoskeletal function among experimental groups of the hiPSCs lines, EZR (EZRIN) turned out to be the top-upregulated protein (**Figure 6**). Ezrin, also known as cytovillin or villin-2, is a cytoplasmic peripheral membrane protein and functions as a substrate for tyrosine kinase in microvilli. Its significance for hPSCs morphology has not been studied. Earlier, in support of our data, EZRIN was demonstrated as one of the most prominent cytoskeletal proteins by proteomic profiling of hESCs at the first 48 hours of the early differentiation stage [12], suggesting that it may be expressed differently in clones with "good" and "bad"

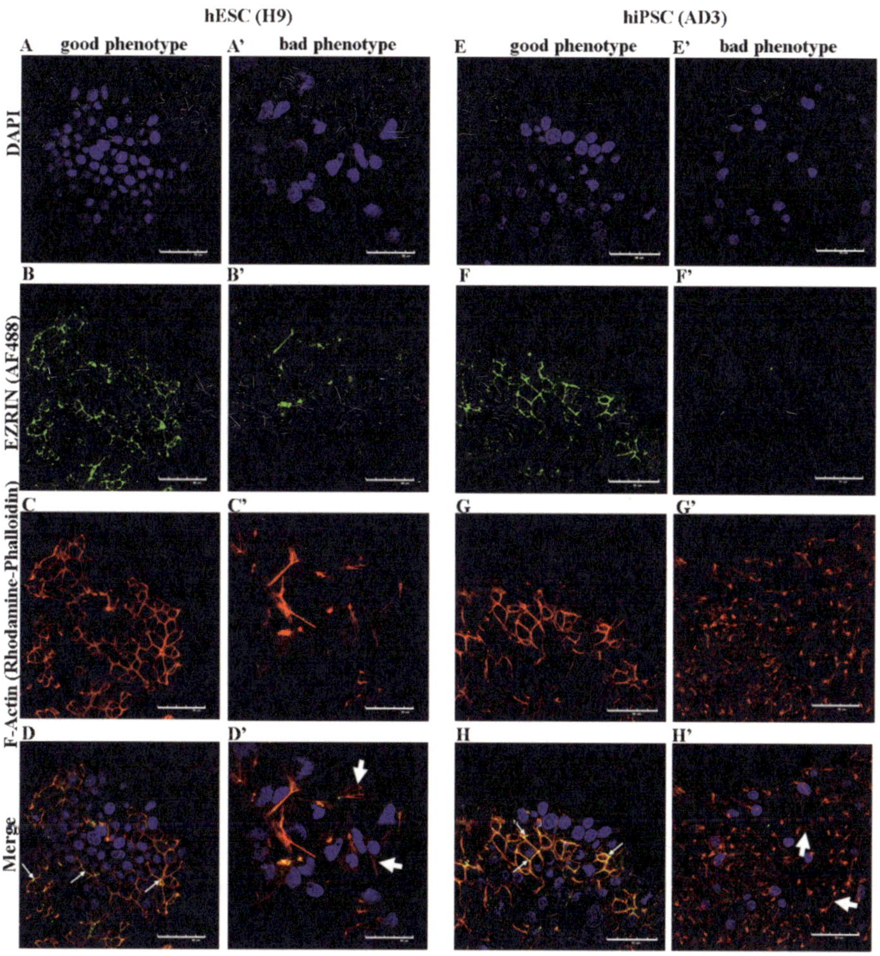

Figure 7.
Association of the EZRIN cytoskeletal network with F-actin in hESC and hiPSC clones with "good" morphological phenotype and complete distraction of this network in "bad" clones. Thin white arrows are pointing on colocalization of F-actin and EZRIN; thick white arrows show the absence of colocalization. Immunofluorescence was performed using anti-EZRIN antibody 3C12 (Invitrogen, 35-7300, secondary), secondary anti-mouse IgG tagged with Alexa Fluor 488 (Abcam, ab150113), and Rhodamine-Phalloidin (Invitrogen, R415). Scale bar 50 μm.

phenotype. In support of our proteome data, by employing specific anti-EZRIN antibody, we were able to detect the association of the EZRIN with F-actin in clones with good morphological phenotype and complete absence of such association and very weak pattern of staining in "bad" clones (**Figure 7**).

Interestingly, a very close in concept earlier work of Bjørlykke and colleagues [14] performed on 20 hiPSC lines different on their morphological appearance did not recognize the same cytoskeleton proteins among abundant upregulated or down-regulated proteins. However, KERATIN 19 (KRT19), a member of the keratin family of the intermediate filament proteins responsible for the structural integrity of the epithelial cells was identified among upregulated ones as well as ADD2, a member of the cytoskeleton-associated proteins (ADDUCINS) that promotes the assembly of the spectrin–actin network [14]. Among abundant downregulated proteins PALLADIN (PALLD) and FIBRONECTIN1 (FN1) along with the MRC2, extracellular matrix remodeling protein, appeared as significantly downregulated [14]. Fibronectins bind cell surfaces and various compounds as collagen, fibrin, and actin. These proteins involved in cell adhesion and maintenance of the cell shape. Palladin as a cytoskeleton protein involved in the organization of the actin network, motility, and adhesion. Importantly, both named proteins have a role in cell morphology. Keeping in mind the importance of the colony-defined edge as meaningful morphological characteristic of a good hPSCs phenotype, one can recognize an importance of MRC2 for the establishment of the hPSCs phenotype as MRC2 is a member of the mannose receptor family proteins and plays a role in the extracellular matrix remodeling.

Importantly, it appears that only five proteins (SFXN1, LAMC1, RAP2B, NUP98, and EEF2) were identified *via* comparison of the "good" morphological hESC H9 samples versus two hiPSC lines with the same characteristics (**Figure 8**). Wherein, H9 samples contained 83 unique proteins and hiPSCs–116, but none of the identified proteins is a cytoskeletal protein.

Eukaryotic translation elongation factor 2 (EEF2), the GTP-binding translation elongation factor family member and an essential factor for protein synthesis appeared upregulated in hiPSCs while was downregulated in hESCs samples with good morphology (**Figures 2** and **6**). EEF2 is known as a positive regulator of apoptosis [46]. In a highly proliferative cells, EEF2 maintains genomic integrity by arresting the cell cycle at G2/M phase in response to ionizing radiation to prevent

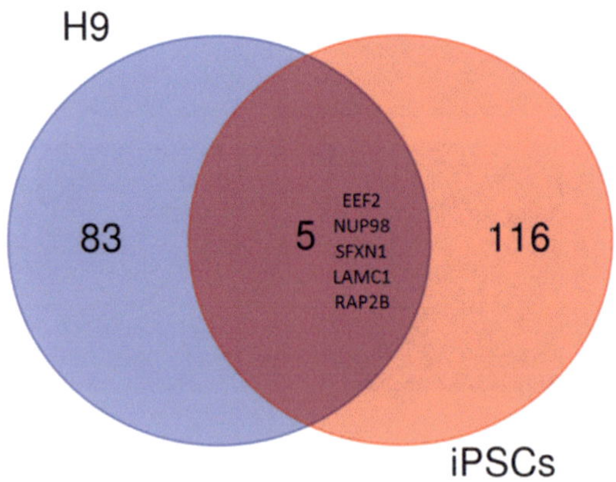

Figure 8.
Comparison of the "good" morphological hESC H9 samples versus two hiPSC lines with the same characteristics revealed five common proteins.

mitotic catastrophe [46]. The rapid proliferation of hPSCs is due to their unique cell cycle regulation. The interplay between cyclins, cyclin-dependent kinase (CDK), and cyclin-dependent kinase inhibitors is important for tight regulation of cell cycle progression in these cells [47]. Moreover, the cell cycle regulation is not only tightly related with pluripotency but also the cell cycle regulators have important functions in DNA damage response (DDR) [42]. Since maintaining genomic stability in hPSCs plays a pivotal role in their self-renewal and stemness, the role of EEF2 should be assessed in the nearest future as in terms of therapeutic application, genomic stability is the key to reducing the risks of cancer development due to abnormal cell replication.

Laminin subunit gamma 1 (LAMC1) belongs to Laminins, a family of extracellular matrix glycoproteins, which are the major non-collagenous constituent of basement membranes. Basement membranes are thin sheets of specialized extracellular matrix (ECM), underlying all epithelia and some other cell types. Laminins are important regulators of cellular functions such as cell adhesion, differentiation, migration, signaling, and metastasis. Human PSCs not only have characteristics typical for epithelial cells [48, 49] but they also rely upon ECM proteins for the support of their niche [50]. Human ESCs produce Laminin α1, α5, β1, and γ1 chains and deposit them as Laminin-511 into hESC-produced ECM. Importantly, Laminin-511 supports hESCs growth in defined medium equally well as Matrigel [50]. Indeed, LAMC1, as well as LAMB1, have been detected in the hESCs by proteomic analysis [11]. However, in our analysis, LAMC1 appears among downregulated proteins in hiPSCs with good phenotype in contrast to hESCs (**Figures 2** and **6**) regardless that all cell lines were grown on the same basement membrane matrix-Matrigel (Corning) with mTESR1 media [6]. As much as hiPSCs are not identical to hESCs, the identified differences may indicate the need for further research in the direction of the hPSCs niche supporting factors for better support of their *in vitro* maintenance.

SFXN1, RAP2B, and NUP98 are expressed in both hESCs and hiPSCs with "good" morphological phenotypes (**Figures 2** and **6**).

Sideroflexin 1(SFXN1) is an integral component of the mitochondrial inner membrane, and it is important for D-serine and L-serine transmembrane transporter activity.

Ras-related protein Rap-2b (RAP2B) is a member of the Ras family of small GTP-binding proteins, and it is involved in innate immune response and ERK signaling, both of which are important players during the reprogramming process. Also, RAP2B may play a role in cytoskeletal rearrangements and may regulate cell spreading through activation of the effector Traf2- and Nck-interacting kinase (TNIK) [51]. Moreover, RAP2B is expressed at high level in various human tumors, where its involvement in cellular spreading and migration was demonstrated more recently [52].

Nuclear pore complex protein Nup98 (NUP98) plays a role in the nuclear pore complex (NPC) assembly and/or maintenance. Nuclear pore complex (NPC) proteins are well-known for their critical roles in regulating nucleocytoplasmic traffic of macromolecules across the nuclear envelope. Several findings suggest that some nucleoporins, including Nup98, have additional functions in developmental gene regulation. Nup98 exhibits transcription-dependent mobility at the NPC but can also bind chromatin away from the nuclear envelope, and it is frequently involved in chromosomal translocations [53]. Importantly, acting as transcription factor, Nup98, could interact directly with histone-modifying enzymes CBP/p300 and histone deacetylases (HDACs), the role of which for hPSCs is well established. However, while the role of Nup98 as a multifunctional protein in macromolecular export has been studied extensively [53], its precise role in hPSCs has not been elucidated and awaits further discovery.

Thus, we can conclude that despite the significant differences in the protein content for the two studied pools (hESCs vs. hiPSCs), when comparing cells of different types within the same experimental group of "good" morphological phenotype, only five differentially expressed proteins were found out of 1933 reliably identified proteins. This may indicate the similarity of the mechanisms that regulate the "good" morphology of hPSCs.

4. Conclusions and future perspectives

The development of reliable methods for estimating the quality of the hPSCs cultures is an urgent requirement for their reliable use in the clinic. Currently, much attention is paid to the creation of the automatic methods for selecting the best clones based on their images, as noninvasive methods for their evaluation. The first section of our chapter is devoted to these methods with a particular emphasis of our approaches [6, 7] based on the analysis of the morphology of colonies and cells. However, the search for the new approaches to analyze morphological parameters should not stop and the question of the regulation of the cell morphology deserves a separate chapter.

Our proteomic data for the first time demonstrated cytoskeletal proteins as top effectors of the "good" morphological hPSCs phenotype. As discussed above, most of these cytoskeletal proteins have not been studied in detail in hPSCs. The molecular differences on the proteome level between hiPSCs and hESCs lines, as reported in multiple publications, may be related to many factors such as time in culture, methods of cells propagation, general culture conditions, as well as different somatic origin of hiPSCs, the level of pluripotency, and many others [10, 54, 55]. Regardless of the used approaches and cell lines, all proteomics results revealed a large proportion of cytoskeletal proteins, thus highlighting cytoskeletal remodeling as a prominent characteristic for hPSCs phenotype [8–10, 12, 13]. That is not surprising, as the actin cytoskeleton network, consisting of actin filaments and crosslinking and motor proteins, regulates the shape of the most cells.

Understanding the mechanisms responsible for the dynamic changes of the colony morphology from the "good" to the "bad" is an important prerequisite for the safe clinical application of these cells, not only because the differentiation potential of hPSCs is deeply associated with the colony morphology but also because the morphological changes occur quicker and well before significant changes in the pluripotency markers expression profiles can be detected [47, 56]. Human PSC colonies demonstrate fast changes of morphological parameters during the exponential growth, and essential differences in their structure associated with the colony area, mean nuclei area, and mean distance between nearest neighbors were shown to be good indicators to detect possible changes of the pluripotency status [57]. So far, we are only making the first steps toward the complete understanding of this process.

Based on our data, we propose to expand the panel of hPSCs markers used to identify the "best" morphology phenotype to include the cytoskeletal proteins, namely MYH7, RDX, and CNN3 for evaluation of the best hESCs and EZRIN for evaluation of hiPSCs. Obviously, the quest for the reliable markers for the identification of the best morphology has to continue.

Eventually, the development of more complex automated approaches for comparative analysis of cells will provide the best quality control of clones, which will thus ensure their continued safe application in regenerative medicine.

Acknowledgements

This study was funded by the Russian Science Foundation, grant number 21-75-20132 for IN. Proteomic analysis was performed using the equipment of the "Human Proteome" Core Facility Center of the Institute of Biomedical Chemistry (Moscow, Russia). The authors express their deep gratitude to Prof. S.L. Kiselev and A.V. Panova for providing the hiPSC line HPCASRi002-A (CaSR), and to Dr. Ivan Diakonov from Imperial College London for valuable comments and suggestions during the paper preparation. We are also thankful to the staff of the Institute of Cytology core microscopic facility M.L. Vorobev and G.I. Stein.

Conflict of interest

The authors declare no conflict of interest.

Author details

Vitaly Gursky[1†], Olga Krasnova[1†], Julia Sopova[1], Anastasia Kovaleva[1], Karina Kulakova[1], Olga Tikhonova[2] and Irina Neganova[1*]

1 Institute of Cytology, Saint Petersburg, Russia

2 Institute of Biomedical Chemistry, Moscow, Russia

*Address all correspondence to: irina.neganova@incras.ru

† These authors contributed equally.

References

[1] Hanna J et al. Direct cell reprogramming is a stochastic process amenable to acceleration. Nature. 2009;**462**(7273):595-601

[2] Deinsberger J, Reisinger D, Weber B. Global trends in clinical trials involving pluripotent stem cells: A systematic multi-database analysis. NPJ Regenerative Medicine. 2020;**5**:15

[3] Chin MH et al. Induced pluripotent stem cells and embryonic stem cells are distinguished by gene expression signatures. Cell Stem Cell. 2009;**5**(1):111-123

[4] Tan Y et al. Comparative study using Raman microspectroscopy reveals spectral signatures of human induced pluripotent cells more closely resemble those from human embryonic stem cells than those from differentiated cells. Analyst. 2012;**137**(19):4509-4515

[5] Kilpinen H et al. Common genetic variation drives molecular heterogeneity in human iPSCs. Nature. 2017;**546**(7658):370-375

[6] Krasnova OA, Gursky VV, Chabina AS, Kulakova KA, Alekseenko LL, Panova AV, et al. Prognostic analysis of human pluripotent stem cells based on their morphological portrait and expression of pluripotent markers. International Journal of Molecular Sciences. 2022;**23**:12902

[7] Mamaeva A, Krasnova O, Khvorova I, Kozlov K, Gursky V, Samsonova M, et al. Quality control of human pluripotent stem cell colonies by computational image analysis using convolutional neural networks. International Journal of Molecular Sciences. 2023;**24**:140

[8] Brill LM et al. Phosphoproteomic analysis of human embryonic stem cells. Cell Stem Cell. 2009;**5**(2):204-213

[9] Van Hoof D et al. Proteomics and human embryonic stem cells. Stem Cell Research. 2008;**1**(3):169-182

[10] Pripuzova NS et al. Development of a protein marker panel for characterization of human induced pluripotent stem cells (hiPSCs) using global quantitative proteome analysis. Stem Cell Research. 2015;**14**(3):323-338

[11] Soteriou D et al. Comparative proteomic analysis of supportive and unsupportive extracellular matrix substrates for human embryonic stem cell maintenance. The Journal of Biological Chemistry. 2013;**288**(26):18716-18731

[12] Novak A et al. Proteomics profiling of human embryonic stem cells in the early differentiation stage. Stem Cell Reviews and Reports. 2012;**8**(1):137-149

[13] Jadaliha M et al. Quantitative proteomic analysis of human embryonic stem cell differentiation by 8-plex iTRAQ labelling. PLoS One. 2012;**7**(6):e38532

[14] Bjorlykke Y et al. Reprogrammed cells display distinct proteomic signatures associated with colony morphology variability. Stem Cells International. 2019;**2019**:8036035

[15] Healy L, Ruban L. Atlas of Human Pluripotent Stem Cells in Culture. New York, NY: Springer US: Imprint: Springer; 2015. p. 1 online resource (XV, 206 pages 285 illustrations, 279 illustrations in color)

[16] Yu J et al. Induced pluripotent stem cell lines derived from human somatic cells. Science. 2007;**318**(5858):1917-1920

[17] Wakui T et al. Method for evaluation of human induced pluripotent stem cell quality using image analysis based on the biological morphology of cells. Journal of Medical Imaging (Bellingham). 2017;**4**(4):044003

[18] Maddah M et al. A system for automated, noninvasive, morphology-based evaluation of induced pluripotent stem cell cultures. Journal of Laboratory Automation. 2014;**19**(5):454-460

[19] Tokunaga K et al. Computational image analysis of colony and nuclear morphology to evaluate human induced pluripotent stem cells. Scientific Reports. 2014;**4**:6996

[20] Kato R et al. Parametric analysis of colony morphology of non-labelled live human pluripotent stem cells for cell quality control. Scientific Reports. 2016;**6**:34009

[21] Nishimura K et al. Live-cell imaging of subcellular structures for quantitative evaluation of pluripotent stem cells. Scientific Reports. 2019;**9**(1):1777

[22] Cho YM et al. Dynamic changes in mitochondrial biogenesis and antioxidant enzymes during the spontaneous differentiation of human embryonic stem cells. Biochemical and Biophysical Research Communications. 2006;**348**(4):1472-1478

[23] Folmes CD et al. Somatic oxidative bioenergetics transitions into pluripotency-dependent glycolysis to facilitate nuclear reprogramming. Cell Metabolism. 2011;**14**(2):264-271

[24] Prigione A et al. The senescence-related mitochondrial/oxidative stress pathway is repressed in human induced pluripotent stem cells. Stem Cells. 2010;**28**(4):721-733

[25] Nishimura K et al. A role for KLF4 in promoting the metabolic shift via TCL1 during induced pluripotent stem cell generation. Stem Cell Reports. 2017;**8**(3):787-801

[26] Harkness L et al. Media composition modulates human embryonic stem cell morphology and may influence preferential lineage differentiation potential. PLoS One. 2019;**14**(3):e0213678

[27] Wakao S et al. Morphologic and gene expression criteria for identifying human induced pluripotent stem cells. PLoS One. 2012;**7**(12):e48677

[28] Joutsijoki H et al. Machine learning approach to automated quality identification of human induced pluripotent stem cell colony images. Computational and Mathematical Methods in Medicine. 2016;**2016**:3091039

[29] Perestrelo T et al. Pluri-IQ: Quantification of embryonic stem cell pluripotency through an image-based analysis software. Stem Cell Reports. 2018;**11**(2):607

[30] Witmer A, Bhanu B. Generative adversarial networks for morphological-temporal classification of stem cell images. Sensors (Basel). 2021;**22**(1):206

[31] Wakui T et al. Predicting reprogramming-related gene expression from cell morphology in human induced pluripotent stem cells. Molecular Biology of the Cell. 2023;**34**(5):ar45

[32] Harkness L et al. Identification of a membrane proteomic signature for human embryonic stem cells independent of culture conditions. Stem Cell Research. 2008;**1**(3):219-227

[33] Phanstiel DH et al. Proteomic and phosphoproteomic comparison of human ES and iPS cells. Nature Methods. 2011;**8**(10):821-827

[34] My I, Di Pasquale E. Genetic cardiomyopathies: The lesson learned from hiPSCs. Journal of Clinical Medicine. 2021;**10**(5):1149

[35] Harb N, Archer TK, Sato N. The Rho-Rock-Myosin signaling axis determines cell-cell integrity of self-renewing pluripotent stem cells. PLoS One. 2008;**3**(8):e3001

[36] Chen G et al. Actin-myosin contractility is responsible for the reduced viability of dissociated human embryonic stem cells. Cell Stem Cell. 2010;**7**(2):240-248

[37] Watanabe K et al. A ROCK inhibitor permits survival of dissociated human embryonic stem cells. Nature Biotechnology. 2007;**25**(6):681-686

[38] Li G et al. Transcriptomic profiling maps anatomically patterned subpopulations among single embryonic cardiac cells. Developmental Cell. 2016;**39**(4):491-507

[39] Sato N et al. A gene family consisting of ezrin, radixin and moesin. Its specific localization at actin filament/plasma membrane association sites. Journal of Cell Science. 1992;**103**(Pt 1):131-143

[40] Sato N et al. Radixin, a barbed end-capping actin-modulating protein, is concentrated at the cleavage furrow during cytokinesis. The Journal of Cell Biology. 1991;**113**(2):321-330

[41] Persson A, Lindberg OR, Kuhn HG. Radixin inhibition decreases adult neural progenitor cell migration and proliferation in vitro and in vivo. Frontiers in Cellular Neuroscience. 2013;**7**:161

[42] Neganova I. The role of cell cycle regulation on reprogramming efficiency. In: Birbrair A, editor. Elsevier Series "Advances in Stem Cell Biology". Cambridge, Massachusetts, U.S.A.: Academic Press; 2020. pp. 1-42

[43] Shibukawa Y et al. Calponin 3 regulates actin cytoskeleton rearrangement in trophoblastic cell fusion. Molecular Biology of the Cell. 2010;**21**(22):3973-3984

[44] Ciuba K et al. Calponin-3 is critical for coordinated contractility of actin stress fibers. Scientific Reports. 2018;**8**(1):17670

[45] Maddala R et al. Calponin-3 deficiency augments contractile activity, plasticity, fibrogenic response and Yap/Taz transcriptional activation in lens epithelial cells and explants. Scientific Reports. 2020;**10**(1):1295

[46] Liao Y et al. Paradoxical roles of elongation factor-2 kinase in stem cell survival. The Journal of Biological Chemistry. 2016;**291**(37):19545-19557

[47] Neganova I et al. Expression and functional analysis of G1 to S regulatory components reveals an important role for CDK2 in cell cycle regulation in human embryonic stem cells. Oncogene. 2009;**28**(1):20-30

[48] Ullmann U et al. Epithelial-mesenchymal transition process in human embryonic stem cells cultured in feeder-free conditions. Molecular Human Reproduction. 2007;**13**(1):21-32

[49] Van Hoof D et al. Feeder-free monolayer cultures of human embryonic stem cells express an epithelial plasma membrane protein profile. Stem Cells. 2008;**26**(11):2777-2781

[50] Vuoristo S et al. Laminin isoforms in human embryonic stem cells: Synthesis,

receptor usage and growth support. Journal of Cellular and Molecular Medicine. 2009;**13**(8B):2622-2633

[51] Fu CA et al. TNIK, a novel member of the germinal center kinase family that activates the c-Jun N-terminal kinase pathway and regulates the cytoskeleton. The Journal of Biological Chemistry. 1999;**274**(43):30729-30737

[52] Di J et al. Rap2B promotes cell proliferation, migration and invasion in prostate cancer. Medical Oncology. 2016;**33**(6):58

[53] Franks TM, Hetzer MW. The role of Nup98 in transcription regulation in healthy and diseased cells. Trends in Cell Biology. 2013;**23**(3):112-117

[54] Benevento M, Munoz J. Role of mass spectrometry-based proteomics in the study of cellular reprogramming and induced pluripotent stem cells. Expert Review of Proteomics. 2012;**9**(4):379-399

[55] Kim SY et al. Comparative proteomic analysis of human somatic cells, induced pluripotent stem cells, and embryonic stem cells. Stem Cells and Development. 2012;**21**(8):1272-1286

[56] Neganova I et al. CDK1 plays an important role in the maintenance of pluripotency and genomic stability in human pluripotent stem cells. Cell Death & Disease. 2014;**5**(11):e1508

[57] Orozco-Fuentes S et al. Quantification of the morphological characteristics of hESC colonies. Scientific Reports. 2019;**9**(1):17569

Chapter 2

Undifferentiated and Differentiated Spermatogonial Stem Cells

Danial Hashemi Karoii and Hossein Azizi

Abstract

Spermatogenesis is initiated and sustained by a rare population of singular spermatogonial stem cells (SSCs). These SSCs are connected to the basement membrane of the seminiferous tubules and possess distinctive morphological characteristics. They serve as a vital foundation for a robust stem cell system within the testis, crucial for spermatogenesis and reproductive processes. The isolation and cultivation of human SSCs would significantly enhance our understanding of germ and stem cell biology in humans. Although a challenging endeavor, the recent advancements in enriching and propagating spermatogonia carrying the male genome offer a significant stride toward future transplantation and the restoration of fertility in clinical settings.

Keywords: spermatogenesis, spermatogonial stem cells, stem cell, transplantation, fertility

1. Introduction

Male reproductive systems are characterized by spermatogenesis, which is a genital process. It is the spermatogonial stem cells (SSCs) that play the most important role in this system [1]. Somatic cells and Sertoli cells support SSCs in the basement membrane of seminiferous tubules. Seminiferous tubules support spermatogenesis by containing Sertoli cells (sustentacular cells) [2]. By secreting growth factors during spermatogenesis, these somatic cells support the fate of SSCs.

In spermatogenesis, genetic information is transmitted to the next generation. This process results in spermatogonia differing and proliferating into spermatozoa [3]. They can be subdivided into single spermatogonia (As), paired spermatogonia (Apr), or aligned spermatogonia (Aal) based on their topological organization [4]. Undifferentiated spermatogonia are called spermatogonia collectively. After proliferating throughout the seminiferous epithelium cycle, these cells become quiescent before they differentiate into A1 spermatogonia. An important subsequent division (A2-B) leads to the separation of differentiated SSCs into primary and secondary spermatocytes. As spermatozoa elongate following several meiotic divisions, secondary spermatocytes become round spermatids. Differentiated spermatogonia A1 through B are collectively referred to as spermatogonia A1 through B [5].

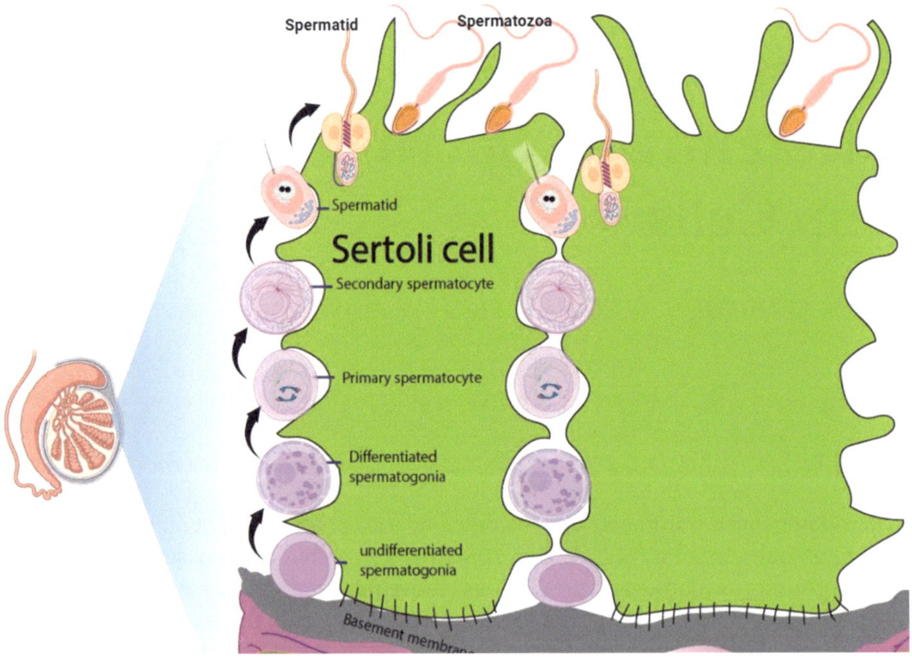

Figure 1.
Spermatogenesis can be divided into five successive stages of germ cell development: (1) spermatogonia, (2) primary spermatocytes, (3) secondary spermatocytes, (4) spermatids, and (5) spermatozoa.

Unipotent stem cells called spermatogonial stem cells (SSC) produce sperm throughout the male's lifetime. In the development, signaling pathway, growth regulation, and differentiation of cells, zinc finger, and BTB domain containing 16 (ZBTB16/PLZF) genes play several important roles. Sperm cells, embryonic stem cells, and pluripotent embryonic stem cells express PLZF [6]. Our study examined the expression of PLZF in testis, stem cells, pluripotent embryonic stem cells, and ES-like cells (**Figure 1**).

There are two types of activities shown by spermatogonia. To maintain the primary pool of stem cells, self-renewal by mitotic divisions is necessary, followed by spermatogenesis, defined as the process of dividing undifferentiated spermatogonia into differentiated ones [7]. As a result of cytoskeletal activity, germ cell movements are also regulated in size and shape. Actin, microtubules, and intermediate filaments are part of the cytoskeleton, which governs the activities of those cells. Vimentin is crucial to these spermatogenic processes because it plays a critical role in the cytoskeleton [8].

2. Some gene expression between pluripotent stem cells and testicular germ cells

2.1 PLZF

The PLZF protein, as detected by immunohistochemistry (IMH), was localized in the differentiated tubular cells of the testis tubule center, as well as in the basal compartment of the seminiferous tubules of the undifferentiated testis [9]. A significant difference was found in PLZF IMH-positive cells in adult testis compared to neonates

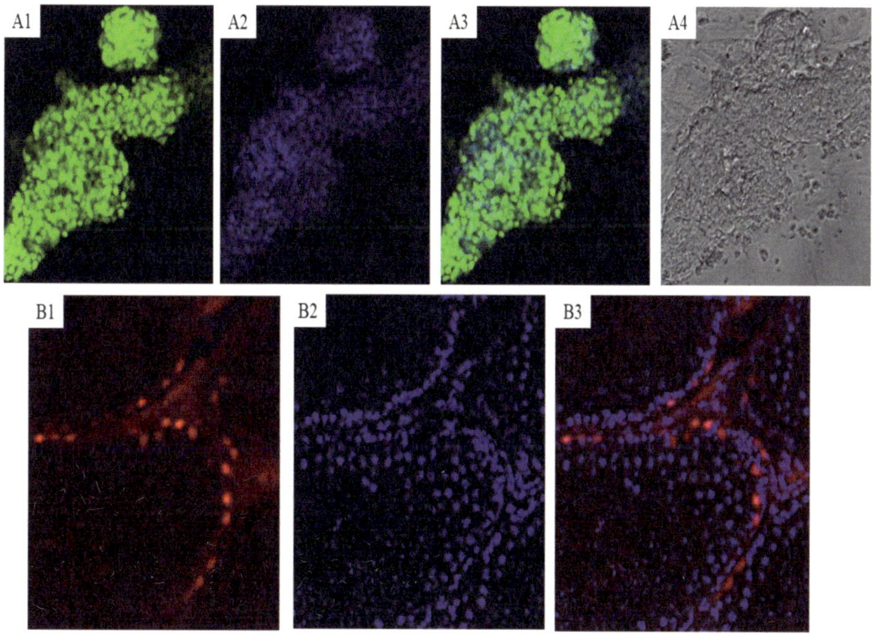

Figure 2.
Testis section with PLZF positive cells. Sections of spermatogonial stem cells from undifferentiated and differentiated spermatogonials (A) and from in vitro spermatogonials (B) were analyzed for PLZF positive cells. There was a greater number of PLZF positive cells in undifferentiated testes compared to differentiated testes and in vitro testes.

when positive cells were counted in sections of seminiferous tubules from undifferentiated and differentiated testes ($P < 0.05$). PLZF germ cell marker was strongly ICC-positive for SSC colonies *in vitro* but negative for ES cells and ES-like cells.

In pluripotent germ cells, PLZF is downregulated, but it is a transcription factor associated with testicular germ cell proliferation. As a result, *in vitro* and *in vivo* analysis of germ cell development can be supported (**Figure 2**).

A marker of spermatogonial differentiation, KIT, is directly repressed by PLZF, according to Filipponi et al. In the testis niche, PLZF plays an important role in maintaining the self-renewal and maintenance of the SSC. In undifferentiated spermatogonia, PLZF is co-expressed with Oct4 [9]. PLZF loss leads to a limited number of normal spermatozoa and, after birth, a lack of respected germline due to the progressive loss of spermatozoa. The expression of genes regulating limb and axial skeletal development is regulated by PLZF during embryogenesis. There is a genetic relationship between PLZF and Gli3 and Hox5 genes during limb development [10]. Testis and SSCs expressed PLZF, making it a SSC marker according to previous studies. PLZF expression in neonate and adult testicular sections, isolated SSCs, ES cells, and ES-like cells of mouse testicular culture was examined to determine if PLZF expression is the same in both testicular germ cells and pluripotent stem cells. According to the results, plunging stem cells do not express PLZF [9].

2.2 Vimentin

There are two types of activity in spermatogonia. To maintain the primary pool of stem cells, self-renewal occurs through mitotic division, followed by spermatogenesis, which refers to the differentiation of undifferentiated spermatogonia into differentiated spermatogonia. Associated with these events are widespread adjustments in germ cell movements in relation to cytoskeletal activity. Microtubules, actin, intermediate filaments, and the cytoskeleton make up the cytoskeleton, which governs the activities of those cells [10–12]. During these spermatogenic processes, vimentin plays a critical role in cytoskeleton function. Spermatogenesis begins with the expression of vimentin, an intermediate filament. The filamentous intermediate filament of vimentin connects the tubulin and actin cytoskeleton to the nuclear periphery [13]. As spermatogenesis progresses, vimentin functions primarily to ensure cellular stiffness, to maintain actin position, to facilitate cell migration, to divide cells, and to organize organelles. Additionally, vimentin has a number of essential roles, including determining cell shape, differentiation, motility, maintaining cell junctions, contributing to the maintenance of ordinary spermatogonia morphology, and anchoring germ cells to the seminiferous epithelium to anchor them [14].

It has been suggested that vimentin plays an important role in the differentiation of SSCs. However, it has been unclear whether the vimentin intermediate filament is necessary during differentiation *in vitro*. Finally, vimentin is expressed in male germ cells in a few studies. Separating germ cells in the adluminal and luminal compartments of seminiferous tubules expressed high levels of vimentin, while undifferentiated cells located in the basal compartment expressed low levels of vimentin. Immunohistochemical analysis indicated that vimentin expression was associated with Sertoli cells near the basal membrane. Afterward, we differentiated germ cells from Sertoli cells using SOX9 specific markers. According to IMH analysis, Sertoli cells expressed SOX9 cytoplasmically and differentiating germ cells expressed SOX9 negatively (**Figures 3** and **4**). Growth factors were added to isolated cells after enzyme digestion. In a previous study, we characterized isolated testicular cells. Immunocytochemistry was used to examine vimentin expression in primary and secondary spermatocytes, round spermatids, and undifferentiated spermatogonia [11].

Using the DAZL specific marker, we distinguished primary and secondary spermatogonia as well as spermatids from undifferentiated spermatogonia. Undifferentiated spermatogonia expressed DAZL at a high level, while differentiated germ cells did not. Differentiating germ cells express high levels of vimentin, while

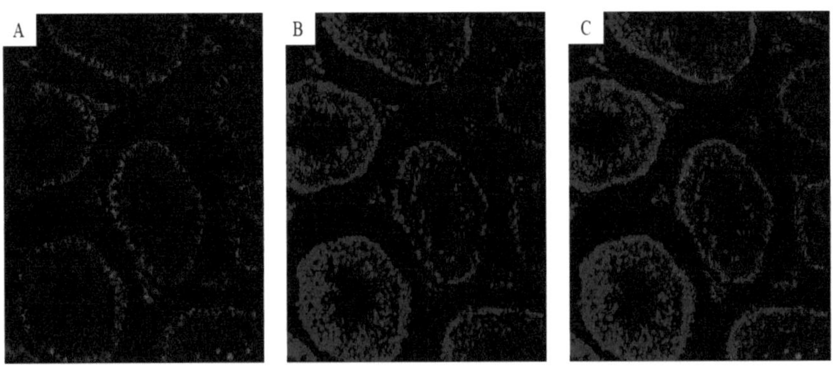

Figure 3.
Characterization of vimentin intermediate filaments in adult mouse seminiferous tubules by immunohistochemistry. Differentiating germ cells located in the middle compartment of seminiferous tubules expressed high levels of vimentin, whereas undifferentiated cells located in the basal compartment expressed low levels. Merged image with DAPI. (A) Vimentin, red, (B) DAPI, blue (scale bar: 15 μm), and merge (C).

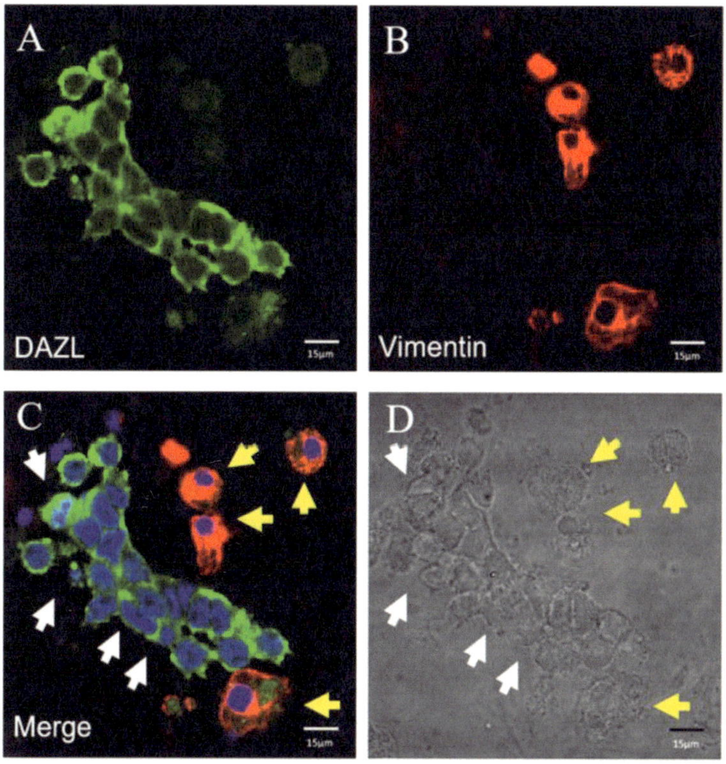

Figure 4.
DAZL and vimentin immunocytochemistry. In (A), DAZL expression is marked by green fluorescence; in (B), vimentin expression is marked by red fluorescence; in (C), DAPI is shown merged with the green fluorescence. The white arrow represents undifferentiated spermatogonia, and the yellow arrow represents germ cells that are differentiating. DAZL is green; vimentin is red; and DAPI is blue (scale bar: 15 μm), and (D) bright field. (scale bar: 15 μm) (get this figure from our recent article [11]).

undifferentiated spermatogonia display low levels. Last time, spermatogonia were differentiated from undifferentiated spermatogonia by utilizing the Ki67 specific marker. As expected, differentiated germ cells express a high level of Ki67, while undifferentiated spermatogonia display a low level of Ki67 expression [11, 14].

2.3 POU5F1

With the advent of SSC culture techniques and genetic analysis, important genes were identified that maintain the stem-cell function of SSCs. As, Apr, and Aal spermatogonia express POU5F1 (POU domain, class 5, transcription factor 1), one of the molecular markers of undifferentiated spermatogonia. The POU5F1 gene encodes a transcription factor that plays an essential role in controlling embryonic development and maintaining pluripotency and self-renewal. POU5F1A, POU5F1B, and POU5F1B1 are produced during alternative splicing, but only POU5F1A maintains stemness.

Developing and optimizing treatment methods for male infertility requires understanding how SSCs differentiate and how genes involved in spermatogenesis are expressed at different stages of SSC differentiation. Since male fertility depends on accurate spermatogenesis and the population of SSCs in the testis, understanding the mechanisms behind SSC differentiation is essential. The POU5F1 protein localization in neonate and adult mice testis did not distinguish the populations of SSCs in our previous study using three antibodies [15]. A comparison between this study and the previous study in this article reveals differences in POU5F1 expression between the two populations of spermatogonia. Since SSCs play an important role in regenerative medicine, it is important to understand the differences between two different populations of SSCs in order to utilize each one appropriately. Our research has helped the scientific community gain a better understanding of what POU5F1 actually does during spermatogenesis [16].

In mouse seminiferous tubules, we analyzed the expression pattern of POU5F1 using immunohistochemistry. Using POU5F1 Proprietary Antibody, we identified SSCs using immunohistochemistry. After merging and staining mouse seminiferous tubules with DAPI, it was found that they express this marker. Seminiferous tubule basal spermatogonial cells exhibit the highest POU5F1 expression, as seen in fluorescent microscope images [17].

Images obtained from the bright field microscope demonstrated the difference between differentiated and undifferentiated spermatogonia when cells were extracted from two spermatogonial populations and cultured on their respective media. Spermatogonial cells that were undifferentiated did not grow much after culturing; they were also small and tended not to form specific shapes. Unlike undifferentiated spermatogonial cells, differentiated spermatogonial cells expand during culturing and tend to form clusters. Thus, spermatogonial cells can be diagnosed based on their morphology [17].

The comparison of two SSC populations shown that there are differentiated and undifferentiated. A bright field of spermatogonia cells during differentiation indicates that undifferentiated and differentiated spermatogonial cells differ morphologically. In the subsequent study, we examined POU5F1 expression in differentiated and undifferentiated spermatogonia by ICC. In two spermatogonia populations, DAPI staining was used to determine SSCs and to stain for the POU5F1 marker. According to ICC analysis of images obtained with a scanning UV laser microscope, undifferentiated cells expressed more POU5F1 than differentiated cells. According to IMH analysis, basement membrane cells expressed POU5F1 at high levels and differentiated cells expressed it at low levels (**Figure 5**).

2.4 VASA

It was discovered that the VASA gene plays a crucial role in the development of female germ stem cells (GSCs) in Drosophila [18]. The VASA gene is eliminated in

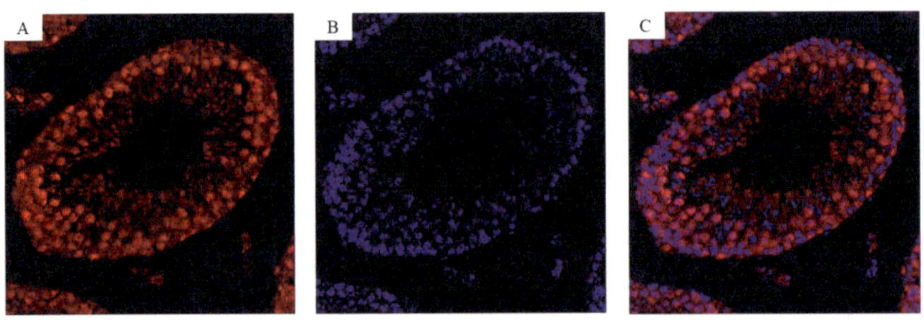

Figure 5.
In the seminiferous tubules, immunohistochemistry (IMH) analysis revealed the expression pattern of POU5F1. POU5F1 (A), DAPI (B), and 4′,6-diamidino-2-phenylindole (POU5F1) (C) show the mixed images and sharp expression of POU5F1 in the basal membrane (scale bar: 15 μm).

mice with a systematic genetic deficiency that results in a loss of sperm production in the males. During meiosis phases, GSCs in males appear to die at the zygotene stage, but the ovary appears to function normally [19]. On embryonic day 12.5 and subsequent to entry into the gonadal anlage, mice show localization of VASA in PGCs. PLZF has been implicated in direct repression of Kit transcription, a spermatogonial differentiation marker, in previous studies. The loss of the PLZF gene also causes limited numbers of normal spermatozoa to be produced, resulting in an impaired germline after birth. In embryogenesis, PLZF regulates gene expression during the patterning of the limbs and axial skeleton. Two types of cell populations present in seminiferous tubules were analyzed for co-expression of PLZF and Oct4.

Spermatogenesis defects are often responsible for infertility in humans. It is essential to understand normal spermatogenesis in order to develop subfertility and infertility in humans. RNA-binding proteins play a crucial role in the formation of germ cells. In addition to rhesus macaques, goats, cattle, pigs, and other animals, VASA is expressed in germ cells [20]. ATP-dependent RNA helicases and RNA-binding proteins are encoded by the VASA gene. Spherical spermatids, spermatogonia, and spermatocytes can be identified in human testicular tissues based on the expression of the VASA protein. Human spermatogenesis might be better understood by understanding how these proteins are expressed in different germ cells at different stages [21, 22].

Drosophila cells dispersed VASA protein evenly throughout their cytoplasm. VASA proteins function as RNA chaperones and are connected to chromatoid bodies. Various studies have shown that VASA functions as the mRNA transcript and CB in spermatozoa when the genome is inactive. In addition to spermatogenesis, VASA is essential for the differentiation of embryonic stem cells into primordial germ cells (**Figure 6**).

2.5 SOX2

In stem cells and progenitor cells, Sox2 plays an important role in maintaining pluripotency and differentiation. Furthermore, it plays a role in cell reprogramming in the inner cell mass (ICM) and ectoderm of blastocysts (10). As well as regulating Sall4, Plzf, Gfra1, Oct4, Klf4, Foxm1, Cux1, Zfp143, Trp53, E2f4, Esrrb, Nfyb, and c-Myc, Sox2 can also convert somatic cells to pluripotent stem cells. Reprogramming and pluripotency are dependent on each other for the production of induced pluripotent stem cells (iPS) [23]. Mice's primordial germ cells also expressed Nanog and Sox2. Human primordial germ cells may still contain Sox2, but further research

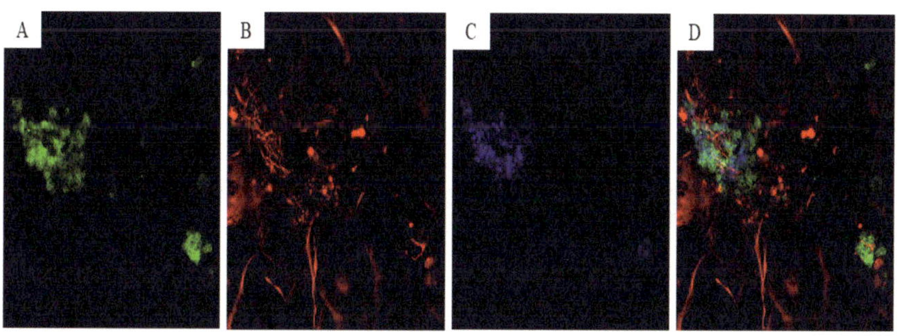

Figure 6.
Immunohistochemy image VASA and vimentin in vitro. (A) Immunohistochemy of VASA, (B) Immunohistochemy image of vimentin, (C) DAPI, and (D) merge (scale bar: 100 μm).

is needed to confirm this. The number of pluripotent cells decreases when Sox2 expression is reduced, and cell differentiation begins. There have also been reports of high levels of Sox2 expression in brain cancers that can cause pituitary tumors and decreased levels of Sox2 expression in patients with ocular abnormalities [24]. Breast cancer, colorectal cancer, and glioblastoma have been linked to increased Sox2 expression, while gastric cancer is associated with its decrease. Different cell types and tissues express Sox2 differently in humans. Bone marrow, endometrium, heart, kidney, liver, and pancreas have lower expression levels than the lung, prostate gland, stomach, testis, and fallopian tube. Sox2 gene interaction with spermatogenesis genes was the goal of this study. We investigated and compared gene expression in differentiated and undifferentiated spermatogonial stem cells. Moreover, this gene was examined in differentiated and undifferentiated spermatogonia to determine its quantity and mode of expression. To improve male infertility treatment, this research aimed to understand the mechanisms involved in sperm generation [25].

The interaction network between Sox2 protein and some other proteins involved in spermatogenesis was analyzed in this experiment, as shown in **Figures 1** and **2**. Using STRING and Cytoscape databases, key genes were identified that are not connected to Sox2, including Sim2 and Rfx4. These two figures illustrate the origins of these genes, the sources of measurement and connection, as well as where each gene is expressed in each testicle and its biological function. A greater or lesser degree of relationship was also detected between genes. Oct4, Nanog, and Klf4 are strongly connected with Sox2, but Smad1, Gdnf, Egr2, and Stra8 are poor connections. In addition to POU5FA, Stra8 and Gdnf, Sox2 appears to be related to Pou5f1, Stra8, Klf4, and Bmp8b. In addition, Sox2 is connected with Pou5f1, Klf4, Kit, and Nanog in stem cell population maintenance. The expression of Sox2 was examined by immunohistochemical analysis of cross sections of seminiferous tubules. According to immunohistochemistry analysis by confocal scanning UV laser microscope, Sox2 nuclear expression increased during spermatogenesis *in vivo* over time (**Figure 3**). In this figure, single undifferentiated cells can be seen along with a group of spermatogonial stem cells. Different expressions of Sox2 were observed in isolated spermatogonia that had been cultured and differentiated *in vitro* (**Figure 4**). The expression of Sox2 in differentiated cells exceeded expectations when considered as a pluripotency factor. A Fluidigm PCR analysis of spermatogonia grown *in vitro* showed that differentiated spermatogonia expressed much more Sox2 than undifferentiated spermatogonia. A high level of Sox2 expression was observed in differentiated cells, and a significant difference in Sox2 expression between differentiated and undifferentiated cells was also observed ($p < 0.05$). Undifferentiated spermatogonia under *in vitro* conditions also expressed high levels of Sox2. However, Sox2 was highly expressed in differentiated spermatogonia, similar to *in vivo* conditions (**Figure 7**). The study's most interesting finding was that undifferentiated cells expressed high levels of Sox2 cytoplasmically and differentiated cells expressed high levels of Sox2 [24].

Sox2 expression is essential for stem cells and SSCs to remain pluripotent, and it also plays an important role in maintaining, increasing, and specializing SSCs and PGCs. Additionally, both *in vivo* and *in vitro* spermatogonia expressed Sox2. The nucleus of differentiated spermatogonia expressed more Sox2 than the cytoplasm of undifferentiated spermatogonia in *in vitro* experiments. Differentiated spermatogonia expressed more Sox2 in *in vivo* experiments than undifferentiated spermatogonia. Still, this gene expression played a critical role in maintaining stem cell pluripotency, which is crucial for spermatogenesis. By studying Sox2 expression in spermatogenesis, we may be able to improve male infertility treatments.

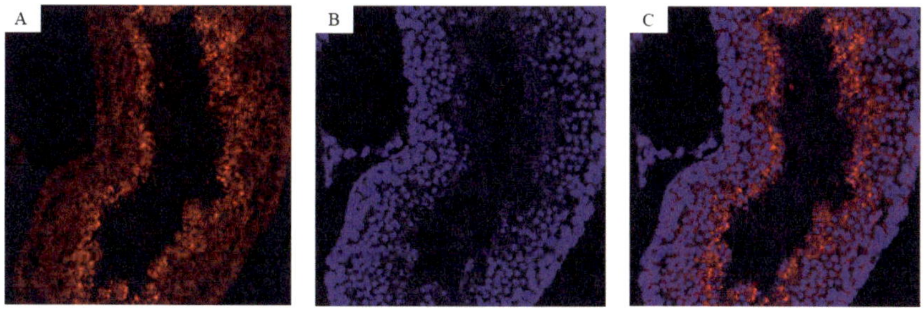

Figure 7.
The pattern of Sox2 expression. (A) the expression of Sox2 in seminiferous tubules, (B) the DAPI (blue) staining shows the nuclear cells, and (C) merged images (scale bar 50 µm).

3. Gene expression profiling of SSCs seems to be age dependent

An embryonic germ layer can be differentiated into ectodermal, mesodermal, and endodermal cells using pluripotent stem cells (PSCs). PSCs have been generated using several different approaches, including ESCs obtained from embryonic blastocysts after fertilization. The so-called induced pluripotent stem cells (iPSCs) were also obtained by enforcing the expression of pluripotency genes in somatic cells; SSCs have proven to be a promising method for establishing PSCs in a more natural and ethically acceptable manner, especially for therapeutic approaches in medicine [26].

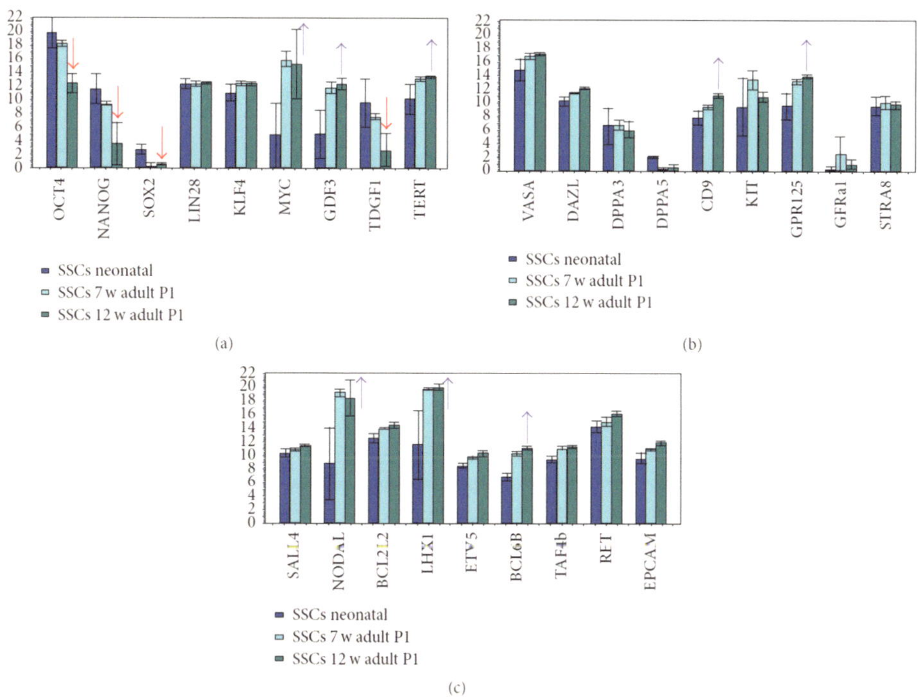

Figure 8.
Neonatal SSCs (colored dark blue), adult SSCs (colored light blue), and adult SSCs (colored blue-green) express pluripotency and germ cell genes differently. The red arrows indicate genes downregulated in adult SSCs, while the purple arrows indicate genes upregulated (more than twofold). Note the downregulation of Oct4, Nanog, and Sox2 in 12-week-old mice's SSCs) (get this figure from our recent article [29]).

It is possible to isolate and expand SSCs *in vitro*, as they are found in small numbers in the testis. They are unipotent stem cells under the control of their stem cell niches, but under specific culture conditions outside the niche and without exogenous pluripotency genes, they are capable of converting into ESC-like cells at various times after culture or isolation of SSCs [27].

In neonatal and adult SSCs obtained from 7- and 12-week-old mice, real-time PCR was used to quantify and analyze the expression of important germ cell-enriched genes (LHX1, Stella, VASA, DAZL, CD9, EPCAM, GPR125, GDF3, THY1, STRA8, GFRa1, 1ITGB1, KIT, ETV5, and BCL6B) and pluripotency associated genes (Oct4, Nanog, Sox2, TDGF4, KLF4, MYC, LIN28, SALL4, and DPPA3).

Figure 7 shows how neonatal SSCs and adult SSCs were grouped according to hierarchical clustering (dendrograms) and principal component analysis (PCA). There was a significant difference between neonatal and adult SSC clusters in the heat map analysis of pluripotency and germ cell genes [28].

The Oct4, NANOG, TDGF1, and Sox2 expression levels of neonatal SSCs were significantly higher than those of adult SSCs. MYC, NODAL, LHX1, GDF3, GPR125, CD9, ITGB1, VASA, TAF4b, EPCAM, BCL2L2, ETV5, DAZL, KLF4, RET, and THY1 were significantly higher expressed in adult SSCs (fold change >2 and -test) than in neonatal SSCs [28].

These differences became even more apparent when neonatal SSCs were compared to SSCs obtained from 12-week-old mice (see Supplementary Tables). In addition, 7-week-old mice have significantly higher expression levels of pluripotency genes than 12-week-old mice (**Figure 8**).

4. An analysis of haGSCs with predefined gene sets related to germline, pluripotency, fibroblasts, and mesenchymal stem cells

There has been some evidence that human adult germ stem cells (haGSCs) derived from highly enriched spermatogonia isolated from adult human testicular tissue are highly versatile and share some similarities with human embryonic stem cells (hESCs). They can be differentiated *in vitro* into a variety of cell lineages comprising the three germ layers and express genes associated with pluripotent cells. Based on some studies, mesenchymal stem cells or cells similar to MSCs may have been the source of cells expressing markers of pluripotency. HaGSCs may also be low-differentiated testicular fibroblasts, according to some studies. The stem cells from human testis biopsy, on the other hand, were derived from both germ and mesenchyme and could differentiate into cells from all three germ layers. Other research has shown that haGSCs can produce small teratomas similar to hESCs. Each of these studies raised new questions regarding the true nature of pluripotency in haGSCs. Generally, the activation of a transcriptional regulatory network is required for the pluripotency of cells, which has been observed in ex vivo cultures of early embryonic cells, as well as germ cells. Members of the pluripotency network are normally active in these cells, including morula and blastocyst-stage (inner cell mass) embryonic cells, epiblasts, primordial germ cells (PGCs), and germline stem cells.

Fluidigm real-time PCR analysis was performed on the following germ cell- and pluripotency-associated genes based on microarray results in addition to the initial panel of germ cell- and pluripotency-associated genes: L1TD1, SALL4, JARID2, HOOK1, EPCAM, PROM1, SALL2, IGFR2BP3, REX1, and GATA4. A similar pattern of gene expression was observed in haGSCs derived from two additional patients, with VASA, DAZL, and PLZF predominant. VASA, DAZL, and PLZF expression in haGSCs was significantly lower than in hSSCs. While STELLA and GFR1 were strongly expressed in haGSCs, the other two germ cell-specific genes were not. As compared to hSSCs, haGSCs expressed REX1, LIFR, and NANOS in similar ranges, while CD9 expressed at a higher level. In hFibs, neither DAZL nor LIFR were expressed. Compared to hFibs, hSSCs and haGSCs showed significantly higher expression of germ cell-associated genes. Similar to hESCs, haGSCs possess a rudimentary gene expression profile associated with germ cells. There were higher levels of CD9 and GFR1 expression in haGSCs than in hESCs (**Figure 9**).

Cell culture produces haGSCs from spermatogonia and MACS enriched in CD49f but never from negatively selected fractions or from patients without spermatogonia. A central cluster of haGSCs with outgrowing "epithelial"-like cells characterized these colonies from hFibs. The expression of germ- and pluripotency-related genes was quite different in haGSCs compared to hFibs based on single-cell Fluidigm analysis. The majority of outliner hESCs and haGSCs did not share any similarities with hFibs.

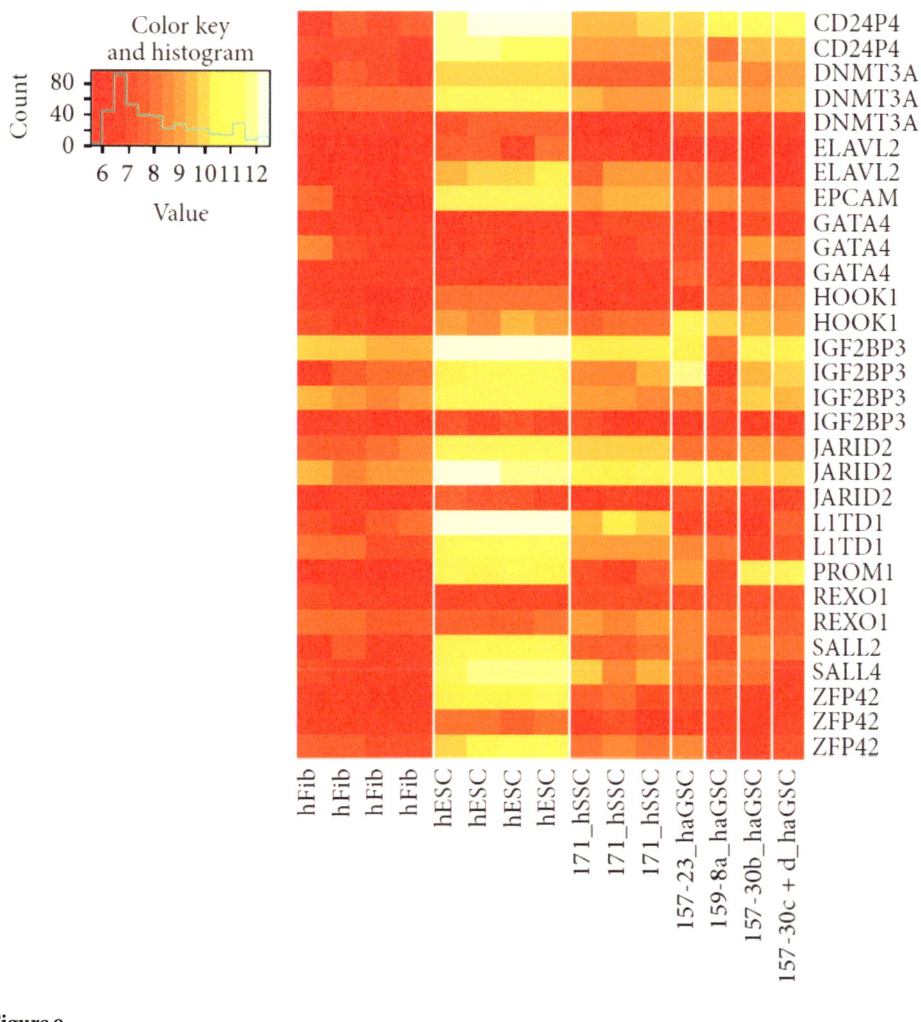

Figure 9.
Based on microarray data, haGSCs are upregulated by pluripotency-associated genes compared to hFibs.

In addition, it became clear that haGSC colonies were heterogeneous, displaying similar characteristics to pluripotent states. Moreover, different haGSC colonies showed a relatively heterogeneous expression of germ- and pluripotency-associated genes in the microarray study. In comparison with hESCs and hFibs, the haGSC transcriptome and high variance genes showed a distinct separation from hFibs.

5. Spermatogonia stem cell gene ontology and signaling pathway bioinformatics analysis

Statistic and bioinformatics analyses are the main bottleneck in transcriptomic studies. Candidates were usually identified using widely accepted statistical criteria (such as P values and fold changes). In order to translate the gene list into biomedical significance, automatic functional annotation was performed using knowledge bases, such as Gene Ontology (GO) and KEGG pathways. We have recently proposed a framework for revising candidate protein lists and identifying novel proteins based on reanalysis of published proteomics data. We also believe that reanalyzing transcriptomes using optimized bioinformatics methods would help us interpret the results better.

A previously published dataset was used to extract the expression data for two cell types (primitive and differentiated type A spermatogonia) from a previously published study. In the next step, eight canonical markers were evaluated using RNA-Seq data. As well as the expression index, we proposed a new parameter for integrating absolute and relative expression abundances. Our statistical model used this parameter to dynamically select the best cutoff considering biological relevance. To understand and study the maintenance of SSCs, we constructed a refined network by combining information about physical interaction, expression change, biological function, and disease association.

Despite transcriptomics' ability to profile gene expression and regulation, bioinformatics analysis is crucial for translating gene lists into functional biomedical applications. The two groups are usually screened using a one-size-fits-all cutoff using statistical inference. Considering both absolute abundance and relative change, we ranked genes using the expression index proposed in this study. By taking well-studied genes associated with SSC self-renewal as a positive reference, we developed a statistical model that dynamically screens for the best cutoff to prioritize candidate genes. Based on predicted genes involved in cell proliferation or differentiation, an optimal cutoff was determined for identifying functionally important genes [29].

SSCs are thought to be proliferating and surviving by activating and silencing various endogenous genes in response to exogenous factors. The mechanism of SSC self-renewal *in vivo* is still poorly understood despite the identification of a few key regulators and signaling pathways. Our transient model for self-renewal versus differentiation is based on SG-A cells (primitive versus differentiated). Based on the expert knowledge-guided and dynamic statistical model described above, we identified 1119 candidate genes with the best enrichment of canonical markers. By combining physical interactions, expression changes, cellular function, and disease associations with these genes, we finally created a refined network [30]. A high quality and relevance of gene prioritization can be seen in this network, which contains five of the eight canonical markers. As well as finding novel regulators of SSC self-renewal, we suggest the refined network could be used to identify target genes for male infertility and testicular cancer treatment [16, 29, 31–33].

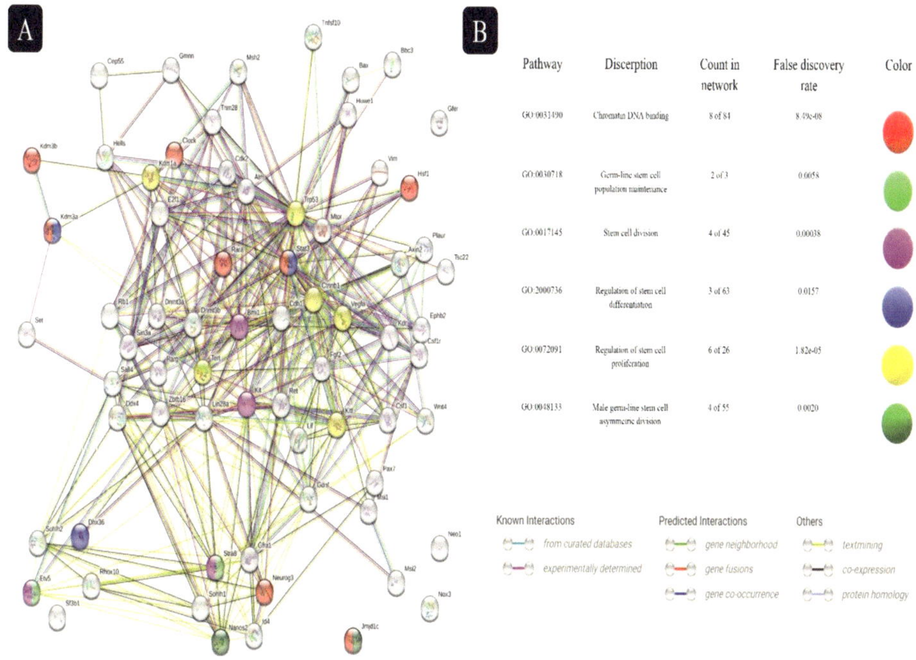

Figure 10.
In silico analysis in spermatogonial stem cell genes. (A) PPI network in spermatogonial stem cell genes, and (B) gene ontology in spermatogonial stem cell.

The protein–protein interaction network with 945 genes was visualized using the STRING (v.11) database. Spermatogenesis was largely regulated by vimentin interaction and regulation, according to the study. There is a strong interaction between vimentin and Stat3, Mmp2, Trp53, Casp7, AURKB, Pik3r1, Ctnnb1, Lgals3, Cdkn1a, and Snai1. There was also a clear association between Trp53, Mmp2, Casp7, Stat3, and Pik3r1. Reactome and KEGG selected any spermatogenesis-related signaling pathway as the master regulator of the pathways involved in spermatogenesis. **Figure 10** shows a strong correlation between the highlighted genes.

6. Conclusion

A large number of spermatogonia are produced during each epithelial cycle when undifferentiated spermatogonia proliferate. When these Aal spermatogonia are in quiescence, they do not divide and develop into the AJ spermatogonia, the first generation of differentiating spermatogonia. Testis undifferentiated regions and the basal section of the seminiferous tubule are strongly expressed with POU5F1, VASA, and PLZF factors, according to the investigation. A comparison of differentiated and undifferentiated populations of spermatogonial stem cells was also conducted. It was found that POU5F1, VASA, and PLZF levels decrease with differentiation, whereas vimentin and sox2 levels increase in differentiated spermatogonial stem cells. In light of the use of SSCs for clinical and therapeutic purposes, such as male infertility, the study of spermatogonial stem cells will provide better insight into the regulation of stem cells in the testis. Also, molecular research and analysis, as well as improved understanding of how genes such as these genes contribute to male infertility, can lead to new treatments or improvements to existing ones. The laboratory can also be used to treat Azoospermia and Oligospermia, abnormal sperm function, and blockages that prevent sperm delivery by investigating the differentiation process of spermatogonial stem cells and better understanding the methods of differentiation.

Acknowledgements

This study was funded by Centre for International Scientific Studies and Collaboration (CISSC), Tehran University and Amol University of special modern technologies.

Conflict of interest

The authors declare no conflict of interest.

Abbreviations

SSCs	spermatogonial stem cells
ZBTB16/PLZF	zinc finger and BTB domain containing 16
ICC	immunocytochemical analysis
iPS	induced pluripotent stem
PSCs	pluripotent stem cells
GO	gene ontology

Author details

Danial Hashemi Karoii[1,2] and Hossein Azizi[2]*

1 Department of Cell and Molecular Biology, School of Biology, College of Science, University of Tehran, Tehran, Iran

2 Faculty of Biotechnology, Amol University of Special Modern Technologies, Amol, Iran

*Address all correspondence to: h.azizi@ausmt.ac.ir

References

[1] Abdelaal NE et al. Cellular therapy via spermatogonial stem cells for treating impaired spermatogenesis, non-obstructive azoospermia. Cell. 2021;**10**(7):1779

[2] Luca G et al. Sertoli cells for cell transplantation: Pre-clinical studies and future perspectives. Andrology. 2018;**6**(3):385-395

[3] Lombó M et al. Genetic and epigenetic alterations induced by bisphenol a exposure during different periods of spermatogenesis: From spermatozoa to the progeny. Scientific Reports. 2019;**9**(1):18029

[4] Hamano KI et al. Spermatogenesis in immature mammals. Reproductive Medicine and Biology. 2007;**6**(3):139-149

[5] Phillips BT, Gassei K, Orwig KE. Spermatogonial stem cell regulation and spermatogenesis. Philosophical Transactions of the Royal Society of London. Series B, Biological Sciences. 2010;**365**(1546):1663-1678

[6] Aponte PM. Spermatogonial stem cells: Current biotechnological advances in reproduction and regenerative medicine. World Journal of Stem Cells. 2015;7(4):669-680

[7] Smith LB, Walker WH. The regulation of spermatogenesis by androgens. Seminars in Cell & Developmental Biology. 2014;**30**:2-13

[8] Hofmann MC, McBeath E. Sertoli cell-germ cell interactions within the niche: Paracrine and Juxtacrine molecular communications. Frontiers in Endocrinology (Lausanne). 2022;**13**:897062

[9] Azizi H, Koruji M, Skutella T. Comparison of PLZF gene expression between pluripotent stem cells and testicular germ cells. Cell Journal. 2020;**22**:60-65

[10] Darzi MB et al. Immunohistochemistry and immunocytochemistry analysis of PLZF and VASA in mice testis during spermatogenesis. Zygote. 2023;**31**(3):273-280

[11] Niazi Tabar A et al. Testicular localization and potential function of vimentin positive cells during Spermatogonial differentiation stages. Animals. 2022;**12**(3):268

[12] Hashemi Karoii D, Azizi H. A review of protein-protein interaction and signaling pathway of vimentin in cell regulation, morphology and cell differentiation in normal cells. Journal of Receptors and Signal Transduction. 2022;**42**(5):512-520

[13] Karoii DH, Azizi H, Amirian M. Signaling pathways and protein–protein interaction of vimentin in invasive and migration cells: A review. Cellular Reprogramming. 2022;**24**(4):165-174

[14] Dunleavy JEM et al. The cytoskeleton in spermatogenesis. Reproduction. 2019;**157**(2):R53-R72

[15] Niknejad P, Azizi H, Sojoudi K. POU5F1 protein and gene expression analysis in neonate and adult mouse testicular germ cells by immunohistochemistry and immunocytochemistry. Cellular Reprogramming. 2021;**23**(6):349-358

[16] Hashemi, Karoii D, Azizi H. OCT4 protein and gene expression analysis in the differentiation of

spermatogonia stem cells into neurons by immunohistochemistry, immunocytochemistry, and bioinformatics analysis. Stem Cell Reviews and Reports. 2023:1828-1844

[17] Masoudi M et al. Comparison of POU5F1 gene expression and protein localization in two differentiated and undifferentiated spermatogonial stem cells. Biologia Futura. 2022;**73**(4):503-512

[18] Yu J et al. Protein synthesis and degradation are essential to regulate germline stem cell homeostasis in drosophila testes. Development. 2016;**143**(16):2930-2945

[19] Abofoul-Azab M et al. Identification of premeiotic, meiotic, and postmeiotic cells in testicular biopsies without sperm from Sertoli cell-only syndrome patients. International Journal of Molecular Sciences. 2019;**20**(3):470

[20] La HM, Hobbs RM. Mechanisms regulating mammalian spermatogenesis and fertility recovery following germ cell depletion. Cellular and Molecular Life Sciences. 2019;**76**:4071-4102

[21] Amirian M et al. VASA protein and gene expression analysis of human non-obstructive azoospermia and normal by immunohistochemistry, immunocytochemistry, and bioinformatics analysis. Scientific Reports. 2022;**12**(1):17259

[22] Mobarak H et al. Amniotic fluid-derived exosomes improved spermatogenesis in a rat model of azoospermia. Life Sciences. 2021;**274**:119336

[23] Reza E, Azizi H. Comparing the expression levels of alkaline phosphatase, Gfra1, Lin28, and Sall4 genes in embryonic stem cells, Spermatogonial stem cells, and embryonic stem-like cells in mice. Journal of Mazandaran University of Medical Sciences. 2022;**32**(210):13-25

[24] Reza E, Azizi H, Skutella T. Sox2 localization during spermatogenesis and its association with other spermatogenesis markers using protein-protein network analysis. Journal of Reproduction & Infertility. 2023;**24**(3):171-180

[25] Reza E, Azizi H, Ahmadi AA. Evaluation and comparison of the expression levels of the ZBTB16 (Plzf) and ZFP genes and alkaline phosphatase in three cell populations: Mouse spermatogonial stem cells, embryonic stem-like cells (Es-like), and embryonic stem cells. Journal of Ilam University of Medical Sciences. 2023;**31**(1):186-193

[26] Ratajczak MZ et al. Hunt for pluripotent stem cell—Regenerative medicine search for almighty cell. Journal of Autoimmunity. 2008;**30**(3):151-162

[27] Vlajković S et al. Possible therapeutic use of spermatogonial stem cells in the treatment of male infertility: A brief overview. Scientific World Journal. 2012;**2012**:374151

[28] Azizi H et al. Derivation of pluripotent cells from mouse SSCs seems to Be age dependent. Stem Cells International. 2016;**2016**:8216312

[29] Wang M et al. Bioinformatics analysis of transcriptomic data reveals refined functional networks for the self-renewal of mouse spermatogonial stem cells. Stem Cells International. 2018;**2018**:5842714

[30] Li S et al. Genetic association and single-cell transcriptome analyses reveal distinct features connecting autoimmunity with cancers. iScience. 2022;**25**(7):104631

[31] Azizi H, Hashemi Karoii D, Skutella T. Whole exome sequencing and In silico analysis of human Sertoli in patients with non-obstructive azoospermia. International Journal of Molecular Sciences. 2022;**23**(20):12570

[32] Hashemi, Karoii D, Azizi H, Skutella T. Altered G-protein transduction protein gene expression in the testis of infertile patients with nonobstructive azoospermia. DNA and Cell Biology. 2023;**12**(8):18-31

[33] Hashemi Karoii D, Azizi H, Skutella T. Microarray and in silico analysis of DNA repair genes between human testis of patients with nonobstructive azoospermia and normal cells. Cell Biochemistry and Function. 2022;**40**(8):865-879

Section 2

Multi-Lineage Differentiation of Human Pluripotent Stem Cells

Chapter 3

Human Pluripotent Stem Cell-Derived Mesenchymal Stem Cells for Oncotherapy

Hao Yu, Xiaonan Yang, Shuang Chen, Xianghong Xu, Zhihai Han, Hui Cai, Zheng Guan and Leisheng Zhang

Abstract

Mesenchymal stem/stromal cells (MSCs) with hematopoietic-supporting and immunoregulatory properties have aroused great expectations in the field of regenerative medicine and the concomitant pathogenesis. However, many obstacles still remain before the large-scale preparation of homogeneous and standardized MSCs with high cellular vitality for clinical purposes ascribe to elusive nature and biofunction of MSCs derived from various adult and fetal sources. Current progress in human pluripotent stem cells (hPSCs), including embryonic stem cells (ESCs) and induced PSCs (iPSCs), have highlighted the feasibility of MSC development and disease remodeling, together with robust MSC generation dispense from the inherent disadvantages of the aforementioned MSCs including ethical and pathogenic risks, donor heterogeneity and invasiveness. Herein, we review the state-of-the-art updates of advances for MSC preparation from hPSCs and multiple tissues (perinatal tissue, adult tissue) as well as tumor intervention with biomaterials, and thus propose a framework for MSCs-based oncotherapy in regenerative medicine. Collectively, we describe the landscape of *in vitro* generation and functional hierarchical organization of hPSC-MSCs, which will supply overwhelming new references for further dissecting MSC-based tissue engineering and disease remodeling.

Keywords: hPSCs, MSCs, drug delivery, oncotherapy, biomaterials

1. Introduction

Human pluripotent stem cells (hPSCs), including human induced PSCs (hiPSCs) and human embryonic stem cells (hESCs), are cell population with unique self-renewal and multi-lineage differentiation potential [1–3]. Attribute to the aforementioned properties, hPSCs have been considered as splendid alternatives for tissue engineering and disease remodeling [3, 4]. For instance, we and other investigators have been devoted to verifying the feasibility of high-efficient generation of MSCs from hPSCs (hPSCS-MSCs) for diverse disease treatment, including osteoarthritis, colitis, liver fibrosis [4–6]. Therewith, hPSCs have served as advantageous alternative sources for MSC preparation for regenerative medicine [7].

MSCs with unique immunoregulatory properties and tissue-repair capacity have been considered as advantageous cytotherapy for various refractory and recurrent disorders. For instance, preclinical studies and clinical practice have suggested the safety and efficiency of MSCs against hematological diseases, articular diseases, neurological diseases, digestive diseases, immune diseases and vascular diseases [8–11]. Meanwhile, the unique characteristic of MSCs with a lower immunogenicity as recommended by the International Society for Cellular Therapy (ISCT), which is appropriate for cell-based cancer immunotherapy [4, 12].

Currently, a certain number of studies have been reported that the capability of MSCs can migrate directionally to tumor sites and contribute to tumor microenvironment formation. Moreover, MSCs exert therapeutic function through an immune evasive mechanism, which will protect MSCs from immune detection and prolong their persistence *in vivo* [13, 14]. Numerous preclinical studies have indicated MSCs as gene transfer systems and ideal drug carries for targeted tumor therapy by releasing cytokines or suppressing tumor cells [15]. For examples, MSCs can load with anti-tumor drugs (as PTX or GBA), enzyme prodrug (as 5-FC/CD, GCV/HSV-TK or CPT-11) or oncolytic viruses, which thus provide antitumor effects with improved safety profiles. In addition, MSCs genetically modified to express interleukin (e.g., IL-2, IL-10, IL-12, IL-15, IL-18, IL-21) and interferon (e.g., IFN-a, IFN-β) could elicit antitumor immunity *in vivo* and inhibit tumor growth in vitro. Although, a large number of pre-clinical studies have been conducted to investigate engineering MSCs and revealed that the effects of it on tumor progress, only a small number of registered and completed clinical trials of engineering MSCs for tumors treatments. In this review, we briefly review the pre-clinical and clinical trials of engineered MSCs as gene transfer systems or drug delivery vehicles for the treatment of solid tumors, as well as summarize the therapeutic mechanism of cancers with engineered MSCs and future prospects.

2. Cell sources for MSC preparation

2.1 Adult tissue-derived MSCs

Since the 1960s, MSCs have been isolated from various sources, including adult tissues (e.g., bone marrow, adipose, dental pulp), perinatal tissues (e.g., umbilical cord, amniotic membrane, placenta) and even derived from human pluripotent stem cells (e.g., hESCs and hiPSCs) [16, 17]. Of them, MSCs were firstly isolated from bone marrow in clinical practice, followed by relative tissues such as adipose tissue, dental pulp and apical root papilla [18]. Bone marrow-derived MSCs (BM-MSCs) have been considered with the widest range of clinical applications, whereas adipose tissue-derive MSCs (AD-MSCs) have been recognized with superiority in adipogenesis over the relative tissue-derived MSCs [4, 19, 20].

2.2 Perinatal tissue-derived MSCs

To date, diverse perinatal tissues have been applied for MSC preparation, including umbilical cord, umbilical cord blood, amniotic membrane, amniotic fluid and placenta. For instance, Zhao *et al.* reported the generation of MSCs from umbilical cord (UC-MSCs) as well as the variations in biological and molecular properties at series passages [12]. Instead, Wei *et al.* and Du *et al.* took advantage of the cytokine cocktail-based strategies for the high-efficient generation of VCAM-1$^+$ UC-MSCs with preferable immunoregulatory and proangiogenic properties [21, 22]. Of note, we and other investigators in the field verified the superiority of UC-MSCs over relative counterpart in immunoregulatory properties [8, 23]. As to placenta tissue-derived MSCs (P-MSCs), Hou *et al.* reported the spatio-temporal metabolokinetics as well as the efficacy upon mice with refractory Crohn's-like enterocutaneous fistula as well [24].

2.3 Human PSCs-derived MSCs

State-of-the-art literatures have reported the generation of MSCs from both hESCs and hiPSCs. Generally, there are four typical procedures for high-efficient hPSC-MSC preparation, including the monolayer model, the coculture model, the embryonic body (EB) model, and the cell programming strategy. For instance, we took advantage of a transcription factor, MSX2, for the initiation of MSC differentiation within 2 weeks [4]. Furthermore, we turned to small molecular cocktail-based strategies for high-efficient hPSC-MSC generation [5]. Notably, the hPSC-MSCs revealed considerable efficacy for the management of colitis, critical limb ischemia (CLI) and osteoarthritis [4–6]. Meanwhile, Li *et al.* and Yan *et al.* reported the therapeutic effect of hESC-MSCs for the treatment of autoimmune and inflammatory diseases under serum-containing or serum-free condition, respectively [25, 26]. Additionally, Wang and the colleagues generated hESC-MSCs with immune modulatory property via a trophoblast-like intermediate stage, which would also help understand the early mesengenesis in vitro [27].

3. Current strategies for MSC engineering

3.1 Nano-engineered mesenchymal stem cells

The therapeutic index of chemotherapeutic drugs can be improved by site-designed administration by reducing the exposure of drugs in non-target tissues. Current methods of targeted drug delivery mainly rely on nano-drug carriers, which can be accumulated in solid tumors. However, this passive accumulation is very inefficient, resulting in less than 5% of the dosage is delivered to the tumor, and the distribution of nano-drug carriers within the tumor is unevenly. More interestingly, MSCs can load with anti-tumor drugs as chemotherapeutic drug paclitaxel (PTX), galbanic acid (GBA) and doxorubicin (DOX), which can uniformly infiltrate into tumor tissue, and improve the distribution of therapeutic drugs within the tumor as shown in **Table 1**. For examples, Pessina *et al.* have demonstrated that MSCs-PTX could produce dramatic antitumor effects in MOLT-4 cells *in vitro* through negatively regulated intercellular adhesion molecule-1 (ICAM-1) and vascular cell adhesion molecule-1 (VCAM-1) expression on microvascular endothelium in 2013 [48]. Moreover, PTX-loaded BM-MSCs and AT-MSCs were respectively co-cultured with NCI-H28 cells and CG5 cells *in vitro*, it showed that both of them could intensely suppress these cancer cell line proliferation and significant improvement in these cell apoptosis [20, 49]. Dual drug loading modalities including cell surface conjugation or endocytosis have been investigated in order to overcome the limited drug loading of MSCs. MSCs have been engineered with various types of organic or inorganic nanoparticles with aimed to improve their drug loading and therapeutic efficacy [35, 50]. For examples, to investigate the efficacy of adipose tissue-derived from MSCs as drug carriers for delivery of galbanic acid

Source of MSCs	Vector systems	Cancer type	Anti-tumor drug	Mainly results	Reference
BM-MSCs	PLGA nanoparticles	Lung cancer	PTX	Incorporating PTX induces upregulation of CXCR4 expression and improves tumor homing	[28]
WJ-MSCs	Exosomes	Cervical cancer	PTX	Incorporating PTX induces apoptosis, and suppressed epithelial-mesenchymal transition proteins in Hela cells	[29]
BM-MSCs	HA-PLGA nanoparticles	Glioma	PTX	The survival of orthotopic glioma-bearing rats was significantly extended	[30]
AD-MSCs	N/A	Ovarian Cancer	PTX	Inhibited ovarian cancer cells migration/ dissemination in 2D and 3D models	[31]
AD-MSCs	N/A	Glioblastoma	PTX	Inhibited the activity of the human pancreatic carcinoma (CFPAC-1) and glioblastoma (U87-MG) by PTX loaded MSCs-TRAIL	[32]
Gingival-MSCs	Exosomes	Pancreatic cancer	PTX	Exerted a significant anticancer effect on both human pancreatic carcinoma and squamous carcinoma cells	[33]
BM-MSCs	Exosomes	Breast cancer	PTX	Decreased the viability of MDA-MB-231 cells *in vitro* and inhibited the tumor growth *in vivo*	[34]
AD-MSCs	PLGA nanoparticles	Colon cancer	GBA	Shown to be efficient in killing C26 colon cancer cells *in vitro* in a dose-dependent manner	[35]
BM-MSCs	Exosomes	Osteosarcoma	DOX	Demonstrates excellent antitumor properties both *in vivo* and *in vitro*	[36]
BM-MSCs	Exosomes	Osteosarcoma	DOX	Shown the low cytotoxicity in myocardial cells and killed the osteosarcoma cells more effectively	[37]
BM-MSCs	Exosomes	Neuroblastoma	DOX	Increased inhibitory effect against NB tumor progression *in vivo* and promote NB cell apoptosis *in vitro*	[38]

Source of MSCs	Vector systems	Cancer type	Anti-tumor drug	Mainly results	Reference
UC-MSCs	Exosomes	Hepatocellular carcinoma	DOX	Cellular uptake and cell cytotoxicity against HepG2 cells *in vitro* and *in vivo*	[39]
BM-MSCs	Exosomes	Osteosarcoma	DOX	Enhanced toxicity against osteosarcoma and less toxicity in heart tissue	[40]
UC-MSCs	Exosomes	Breast cancer	DOX/ CBD	Reduced tumor burden in MDA-MB-231 xenograft tumor model	[41]
BM-MSCs	silica nanoparticles	Hepatocellular carcinoma	DOX	Inhibited the growth of tumors and decreased the side effects in HepG2 xenograft mice	[42]
BM-MSCs	Fe_3O_4 nanoparticle	Osteosarcoma	DOX/ MLT	Improved anticancer efficacy in Saos-2 and MG-63 cells and thus reduced toxicity in normal cells.	[43]
BM-MSCs	Exosomes	Colorectal cancer	DOX	Suppressed C26-tumor growth in vivo	[44]
BM-MSCs	Superpara-magnetic iron oxide (SPIO) nanoparticles	Colon cancer	DOX	Enhanced tumor treatment efficacy of MC38 tumor-bearing C57BL/6 mice	[45]
BM-MSCs	Exosomes	Breast cancer	DOX	Reduced the tumor growth rate of murine breast cancer model	[46]
BM-MSCs	N/A	Breast/thyroid cancer	DOX	Showed enhanced anti-tumor effects in cancer xenograft models	[47]

Table 1.
Pre-clinical experiments of nano-engineered MSCs for cancer therapy within 5 years (2018–2023).

(GBA)-loaded poly (lactic-co-glycolic acid) (PLGA) nanoparticles (nano-engineered MSCs) against tumor cells, the results have performed the nano-engineered MSCs could effectively induce cell death in C26 cells, which is considered to be as a valuable platform for drug delivery in cancer therapy [35]. Remarkably, exosomes derived from MSCs can delivery chemotherapeutic agents (DOX) in the treatment of various cancer. For instance, Liu Y *et al*. have indicated doxorubicin-loaded MSCs encapsulated into superparamagnetic iron oxide (SPIO) nanoparticles could mainly enhance anti-tumor effects and reduce the immune system response in the treatment of colon cancer [45].

3.2 Genetically modified MSCs via non-viral and viral vector systems

During previous years, cytokine-mediated cancer therapy has the potential to enhance immunotherapeutic approaches through the endowing of the immune

Source of MSCs	Vector systems	Cancer type	Cytokine	Mainly results	Reference
UC-MSCs	Lentiviral	Lung cancer	IFN-β	Inhibited the growth of tumor in A549 lung cancer-bearing mice	[52]
G-MSCs	Lentiviral	Squamous cell carcinoma (SCC)	IFN-β	Inhibited the proliferation of tongue squamous cell carcinoma cells *in vitro* and *in vivo*	[53]
AF-MSCs	Non-viral	Lung cancer	IFN-β/ IFN-γ	IFN-primed AFMSCs in suppressing tumor progression *in vivo*	[54]
AD-MSCs	Non-viral	Hepatocellular Carcinoma Cells (HCCs)	IFN-β/ TRAIL	Suppressed proliferation of HCCs through activated STAT1-mediated p53/ p21 by IFN-β, but not TRAIL	[55]
BM-MSCs	Lentiviral	lymphoma	IFN-β/ TRAIL	Exhibited tumor size reduction, growth delay, or apparent tumor clearance	[56]
AD-MSCs	Non-viral	Lung cancer	IFN-β/ TRAIL	reduced tumor weight in H460-derived cancer animal models	[57]
BM-MSCs	Non-viral	Breast Cancer	IFN-γ	increased the apoptosis of MCF-7 cells	[58]
AD-MSCs	Lentiviral	Breast Cancer	IL-2	induced apoptosis in breast cancer cells and stimulated the proliferation of immune cells	[59]
AD-MSCs	Lentiviral	Neuroblastoma	IL-2	Reduced SH-SY5Y proliferation and activate PBMCs *in vitro*	[60]
BM-MSCs	Lipofectamine	Pancreatic Cancer	IL-10	impeded the pancreatic cancer cells proliferation *in vitro* and reduced the growth of tumor xenograft *in vivo*	[61]
BM-MSCs	Lentiviral	Glioblastoma	IL-12	showed a strong inhibitory effect in glioma-bearing nude mice	[62]
BM-MSCs	Lentiviral	Lymphoma	IL-12/ TRAIL	reduced tumor volume and increased survival in mice	[63]
BM-MSCs	Adenovirus	Melanomas	IL-12	inhibition of tumor growth and reduction in the number of metastases in mice	[64]

Source of MSCs	Vector systems	Cancer type	Cytokine	Mainly results	Reference
BM-MSCs	Lentiviral	peritoneal cancer	IL-12/ IL-21	reducing the risk for systemic immune-mediated toxicities	[65]
UC-MSCs	Adenovirus	Glioblastoma	IL-15	exerted stronger therapeutic effects and promoted macrophage/ microglia infiltration in a Vivo model.	[66]
GC-MSCs	Non-viral	Gastric cancer	IL-15	promote tumor cell EMT and induce Tregs ratio increase to affect GC progression	[67]
UC-MSCs	Lentiviral	Breast cancer	IL-18	inhibit the proliferation and metastasis of breast cancer cells *in vivo*	[68]

G-MSCs: gingiva-derived mesenchymal stromal cells; AF-MSCs: amniotic fluid-derived mesenchymal stem cells; and GC-MSCs: gastric cancer-derived mesenchymal stem cells.

Table 2.
Pre-clinical experiments of genetically modified MSCs for cancer therapy within 5 years (2018–2023).

system by providing improved anti-cancer immunity. Nevertheless, the influence of interleukins originated therapeutics is still restricted by short half-life, systemic dose-limiting toxicities, and side-effects. In order to overcome these defects, as gene delivery platform, MSCs have been genetically modified by using viral and non-viral vectors result in the secretion of proinflammatory cytokines to enhance the host immune response to cancer cells, as well as to directly mediate tumor cell death, which have already been reported in several preclinical and clinical trials [51]. Hererin, we have summarized several cytokines engineered MSCs as drug vehicles in the treatment of cancers as seen in **Table 2**.

IFN-β is known to exhibit the classic antitumor effect, which has been certified to inhibit the proliferation of tumor cells and induce apoptosis *in vitro*, however, IFN-β could not generate and maintain therapeutic dose in the tumor sites due to its short half-life; Meanwhile, it leads to the toxicity of organ with serious side effects [52]. To overcome this problem, mesenchymal stem cells (MSCs) have been utilized as drug carriers for IFN-β gene delivery. This IFN-β expressing MSCs as therapeutic agents via systemic administration have been demonstrated effective in attenuation of cancers as melanoma [69], breast cancer [70], pancreatic cancer [71], lung cancer [52], squamous cell carcinoma [53].

IFN-γ can not only enhance the antigen presentation of dendritic cells, up-regulate co-stimulatory molecules, and promote lymphocyte differentiation, and effectively stimulate the activation of effector cells in immune system. Although IFN-γ has many advantages, the ability to induce apoptosis and inhibit angiogenesis will also influence on the normal tissues of body, resulting in side effects. In clinical trials, large doses of IFN-γ have been found to cause the side effects of nervous, blood and liver system. However, using MSCs as a drug carrier with chemotropism and precisely delivery characters, which can not only improve the concentration of IFN-γ in tumor tissues and achieve better therapeutic effectiveness, but also significantly reduce the side effects of IFN-γ on normal tissues.

IL-2 as an immunomodulatory agent was firstly approved by the U.S. Food and Drug Administration (FDA) for the treatment of melanoma and carcinoma, which is required by both effector T lymphocyte and regulatory T cell. However, the short half-life and high-dose toxicity caused by IL-2 limit the clinical application [72, 73]. For instance, Joonbeom Bae and the colleagues reported that exogenous IL-2 gene modified mesenchymal stem cells elicited antitumor immunity and rejuvenate $CD8^+$ tumor-infiltrating lymphocytes (TILs) [74].

IL-10 is produced by innate and adaptive immune cells, and mainly functions as an immune suppressor that inhibits the cancer immunity cycle. However, the half-life of IL-10 in the body is very short. For example, Zhao *et al.* verified that IL-10 modified MSCs could inhibit the growth of the transplanted tumor *in vivo* and prolong survival of bearing animals [61].

IL-12 is mainly produced by antigen-presenting cells (APCs) that regulate the immune response and serves as an effective inducer for T lymphocytes and NK cells to produce interferon-γ (IFN-γ), which is a promising therapeutic agent for the treatment of cancers. However, a short half-life and dose-limited toxicity of IL-12 limits its clinical application [75]. Numerous studies have reported that IL-12 gene modified MSCs could exhibit strengthen the anti-tumor effect in various cancer. For examples, Wu *et al.* have demonstrated that IL-12 derived from lentivirus-mediated IL-12-modified BM-MSCs combined with Fuzheng Yiliu decoction shows a strong inhibitory effect against tumor growth of glioma-nude mice, which have shown promise as an excellent drug delivery vehicle for antitumor-targeted therapy [62]. In another research, Ryu *et al.* took advantage of a delivery system based on IL-12-expressing human umbilical cord blood-derived MSCs (UC-MSCs) significantly inhibited tumor growth and prolonged the survival of glioma-bearing mice, which thus induced long-term antitumor immunity against intracranial gliomas [76].

IL-15 is mainly secreted by activated myeloid cells that are structurally and functionally similar to IL-2. IL-15 supports the persistence of $CD8^+$ memory T cells, while inhibits IL-2-induced T cell death that better maintains long-term anti-tumor immunity [77]. For instance, Wei *et al.* have demonstrated that umbilical cord blood derived MSCs (UCB-MSCs)-transduced with lentivirus vector coding IL-15 could significantly inhibit tumor growth and prolong the survival of Pan02 pancreatic tumor mice, which were associated with tumor cell apoptosis, natural killer (NK) cell—and T-cell accumulation [78].

IL-18 as an interferon (IFN)-γ-inducing factor, which has been reported to be involved in Th1- and Th2- mediated immune responses, as well as in the activation of NK cells and macrophages. IL-18 plays a pivotal role in linking inflammatory immune responses, tumor progression and macrophage activation [79, 80]. For instance, Liu *et al.* indicated that UC-MSCs genetically modified with IL-18 could inhibit the proliferation and metastasis of breast cancer cells *in vivo* by activating immunocytes and immune cytokines, and inhibiting tumor angiogenesis [68].

IL-21 has been reported to induce a cell mediated immune responses, including NK cells and T cells. Moreover, IL-21 as an immunotherapeutic agent has been extensively applied for tumor administration. For examples, Kim *et al.* found that IL-21-expressing MSCs could inhibit the development of disseminated B-cell lymphoma and prolonged survival, which were associated with the infusion of IL-21/MSCs led to induction of effector T and NK cells [81].

4. Programmed MSCs for cytotherapy

4.1 Gene-directed enzyme prodrug therapy

Gene-directed enzyme prodrug therapy (GDEPT) is a novel approach to cancer treatment. Genetically engineered MSCs expressing suicide genes (cytosine deaminase, thymidine kinase, and carboxylesterase) have been indicated to have significant anti-tumor responses as shown in **Table 3**. To date, there are three common pro-drug activating enzymes to modify MSCs (including herpes simplex virus-hymidine kinase (HSV-TK), cytosine deaminase (CD), and rCE) to combine with ganciclovir (GCV), 5-fluorocytosine (5-FC), or Irinotecan hydrochloride (CPT-11), which can effectively inhibit DNA synthesis of tumor, as well as decrease systemic toxicity [86]. As to CD/5-FC, a certain number of researchers have reported MSCs with CD suicide gene expression have been conformed to suppress the development of breast cancer, glioma, melanoma, osteosarcoma and lung carcinoma via converting non-toxic prodrug 5-FC into cytotoxic chemotherapeutic drug 5-FU [92–96]. For instance, Daniela Klimova *et al.* have demonstrated that intravenous injection of adipose-tissue and BM-MSCs-CD/5FC inhibited the progression of tumor in the transgenic adenocarcinoma of the mouse prostate (TRAMP) model [97]. The authors have proposed that MSC/CD combined with 5-FC and TMZ could increase cell cycle arrest and DNA breakage, which could be used in patients with glioblastoma multiforme (GBM) during the immediate postoperative period to sensitize tumors to subsequent adjuvant chemo- and radiotherapy [98]. Moreover, it has been suggested that extracellular vesicles derived from MSCs with CD gene delivery as cargo have an inhibitory effect on the growth of tumor cell lines *in vitro*, as Daniela Klimova engineered the MSCs-EV were cultivated with gemcitabine (GCB), which significantly inhibited the cell growth of pancreatic carcinoma cell lines *in vitro* via converting non-toxic prodrug 5-fluorocytosine (5-FC) to highly cytotoxic prodrug 5-fluorouracil (5-FU), and thereby provide a therapeutic option for tumors [82]. In addition, the transduced iPSC-MSCs both limited growth of preformed tumors and decreased lung metastases after administration of the prodrug (5-FC) [99]. As HSV-TK/GCV, the thymidine kinase (TK)/ganciclovir (GCV) system is a gene-directed enzyme prodrug therapy. Therefore, the herpes simplex virus 1 thymidine kinase (HSV-TK) gene as a suicide gene is introduced into cells phosphorylates a prodrug GCV, which inhibits DNA synthesis and causes cell apoptosis. Although the group of HSV-TK/GCV as suicide gene therapy method is safe and effective in pre-clinical experiments, yet it is not effective in clinical trials due to the lower transfection rate of target cells [100]. In this regard, using engineered MSCs as drug carriers to induce tumor regression in human tumors mainly based on the strong migration ability to especially invasive tumors. For examples, Wei *et al.* have reported that HSV-TK-expressing UC-MSCs combined with prodrug GCV exerted a better effect in the treatment of subcutaneous tumor models and brain intracranial tumor models [88]. Azra Kenarkoohi *et al.* further investigated the anti-tumor activity of MSCs transduced with the HSV/TK in a mouse cervical cancer model via intratumoral injection, which performed significant reduction in tumor size and improvement of NK and CTL activity [85]. As rCE/CPT-11, carboxylesterases (CEs) are enzymes that can convert the prodrug CPT-11 (irinotecan) to its active metabolite SN-38, which has significant cytotoxicity to tumor cells [101]. For example, Seung Ah Choi *et al.* reported that adipose tissue-derived from MSCs

Source of MSCs	Vector systems	Cancer type	Pro-drug	Mainly results	Reference
DP-MSCs	Exosomes	Pancreatic carcinoma	5-FC	Significantly inhibited the cell growth of pancreatic carcinoma cell lines *in vitro*	[82]
AD/UC/ DP-MSCs	Exosomes	Glioblastomas	GCV	inhibited the growth of cerebral C6 glioblastomas *in vivo*.	[83]
AD-MSCs	Microparticles/ ECM	Prostate cancer	GCV	inhibited tumor growth of human prostate cancer *in vivo*	[84]
AD-MSCs	Lentiviral	Cervical cancer	GCV	Significant reduction in tumor size *in vivo*	[85]
AD/BM/ DP/UC/ BP-MSCs	Exosomes	Glioblastoma	GCV	Induce tumor cell death	[86]
BM-MSCs	N/A	Glioblastoma	GCV	Provide a significant growth inhibition and increase survival in a glioblastoma model	[87]
UC-MSCs	N/A	Glioblastoma	GCV	exerts a strong bystander effect on tumor cells	[88]
P-MSCs	Lentiviral	colon cancer	GCV	inhibiting tumor proliferation and inducing tumor apoptosis	[89]
BM-MSCs	PEI-PLL	Glioblastoma	GCV	reduced cell proliferation and angiogenesis in rat C6 glioma	[90]
AD-MSCs	Plasmid	Ovarian Cancer	CPT-11	overcoming drug resistance in ovarian cancer	[91]

DP-MSCs: dental pulp MSCs; AD: Adipose tissue; BM: bone marrow; DP: dental pulp; UC: umbilical cord; BP: blood platelets; P-MSCs: placenta MSCs; and PEI-PLL: polylysine-modified polyethylenimine copolymer.

Table 3.
Pre-clinical experiments of MSCs-based enzyme prodrug for cancer therapy within 5 years (2018–2023).

expressing rCE as cellular vehicles could convert CPT-11 to SN-38, which revealed cytotoxic effect on F98 cell *in vitro* and effectively inhibited the progression of tumor in a rat brainstem glioma model. Therewith, the genetically modified MSCs-rCE as drug delivery have showed therapeutic potential against brainstem gliomas [102].

4.2 Trail prodrug therapy

The death ligand tumor necrosis factor (TNF)-related apoptosis inducing ligand (TRAIL), a member of the TNF cytokine superfamily, has long been recognized for its potential as a cancer therapeutic due to its capacity to induce apoptosis in many types of cancer cells via the receptors DR4 (TRAIL-R1) and DR5 (TRAIL-R2/KILLER), and Fas ligand (FasL) binding to the Fas receptor [103, 104]. Based on the previous research, TRAIL-MSCs as delivery vehicles could induce strength cytotoxicity against cancer cells, which furtherly inhibited tumor growth and prolonged survival in cancer

Source of MSCs	**Vector systems**	**Cancer type**	**Pro-drug**	**Mainly results**	**Reference**
AD-MSCs	Lentiviral	Breast cancer	TRAIL	induce TRAIL-mediated apoptosis *in vitro* and *in vivo* in breast cancer mouse models	[105]
UC-MSCs	Lentiviral	B-ALL	TRAIL	inhibit B-ALL cells proliferation *in vitro* and *in vivo*	[106]
UC-MSCs	Lentiviral	AML	TRAIL/IFN-γ	induce apoptosis in both primary AML patient-derived leukemic cells and AML cell lines	[107]
AD-MSCs	Plasmid	Lung cancer	TRAIL	inhibitory effects on H460 tumor growth both *in vitro* and *in vivo*	[108]
BM-MSCs	Adenoviral	Glioblastoma	TRAIL/VPA	increases the therapeutic effects of MSCs-TRAIL against glioma *in vitro* and *in vivo*	[109]
AD-MSCs	AAV	Hepatocellular carcinoma	TRAIL	inhibit tumor growth and the metastasis of implanted HCC tumors	[110]
BM-MSCs	Exosomes	Hepatocellular carcinoma	TRAIL	enhanced the apoptotic effect of HCC cells *in vitro* and *in vivo*	[111]
BM-MSCs	Plasmid	Melanoma	TRAIL/PEI	induce cell death in B16F0 cells *in vitro* and efficiently reduce tumor weights	[112]
AD-MSCs	N/A	Lung cancer	TRAIL	Protect A549 cancer cells from undergoing apoptosis and increase the survival of cancer cells.	[113]
UC-MSCs	Plasmid	Glioblastoma	TRAIL	significantly higher inhibitory effect and tumor killing effect of gliomas cells *in vitro and in vivo*	[114]
AD-MSCs	Plasmid	Glioblastoma	TRAIL/ Panobinostat	induced decreases in tumor volume and prolonged survival	[115]
BM-MSCs	Adenoviral	Intracranial glioma	TRAIL/VPA	increased migratory capacity toward tumor sites	[109]

B-All: B-cell acute lymphocytic leukemia; AML: acute myeloid leukemia; VPA: valproic acid; AAV: adeno-associated virus; PEI: polyethylenimine; and VPA: valproic acid.

Table 4.
Pre-clinical experiments of TRAIL-MSCs for cancer therapy within 5 years (2018–2023).

models as shown in **Tables 2** and **4**. For instance, Young Un Choi *et al.* constructed the genetically engineered AD-MSCs with TRAIL expression and verified the suppressive effects upon tumor growth in an H460 xenograft model [108]. Chen *et al.* found that TRAIL-MSCs could significantly inhibit the proliferation and promote the apoptosis of B-cell acute lymphocytic leukemia (B-ALL) cells *in vitro* and *in vivo* [106]. Moreover, iPSC-MSCs overexpressing TRAIL are also considered an effective option for the treatment of cancer. For example, Wang and the colleagues have reported that genetically modified iPSCs-MSCs with TRAIL could significantly induce apoptosis in various tumor cell lines *in vitro*, as well as inhibit tumor growth in tumor-bearing mice models via the activation of apoptosis-associated signaling pathways [116].

5. Clinical application of engineering MSCs in tumor

Although numerous preclinical trials have been published, only a small number of clinical trials were registered and completed for the treatment of solid tumors with engineering MSCs. For example, Hanno Niess *et al.* conducted a single-arm phase I/II study for the treatment of gastrointestinal tumors by genetically modified autologous BM-MSCs. According to another clinical trial in the stage of phase I, the safety of the investigational medicinal product (IMP) is evaluated in six patients by 3 times injection of MSCs at diverse concentrations followed by administration of the prodrug Ganciclovir. In the stage of phase II, 16 patients will be enrolled receiving IMP treatment [117]. One completed clinical trial is an investigational study for INF-β modified MSCs in the treatment of ovarian cancer with the aim to evaluated the safety of MSCs/INF-β in the stage of Phase I (without published results). For the treatment of lung cancer, TRAIL engineered allogeneic MSCs as therapeutic agent to treat the metastatic non-small cell lung cancer (NSCLC) patients in a Phase I/II clinical trial. Furthermore, an exploratory trial reported four children with metastatic neuroblastoma to received autologous MSCs infected with ICOVIR-5, and the results exhibited a well-tolerance and safety of MSCs delivered with oncolytic adenoviruses in the treatment of metastatic neuroblastoma [118].

In summary, according to preclinical investigations and clinical trials, we suppose that engineered MSCs as drug delivery is a multifaceted player in oncotherapy development and the clinical transformation of MSCs is urgently needed to accelerate tumor therapy.

6. Prospective and challenges

Longitudinal studies have indicated hPSCs as advantageous cell sources for functional cell generation and the concomitant therapeutic strategy for regenerative medicine and oncotherapy. As mentioned above, the unique property, including self-renewal and multipotent differentiation, have endowed hPSCs with first-rate potential for disease remodeling and alternative cell source preparation. Even though, the significant disadvantages such as teratoma formation and the low differentiation efficiency should not be neglected [3]. Distinguish from the other counterparts, hPSC-MSCs revealed more robust cellular viability and considerable therapeutic effect upon diverse diseases, which thus hold promising prospects for serving as alternative sources of adult tissue- or perinatal tissue-derived MSCs [4].

Notably, considering the rapid progress in gene-editing and MSC-based cytotherapy, it would be of great interesting to further explore the feasibility of generating hESC-MSCs or hiPSC-MSCs with specific targets for the next-generation of oncotherapy in preclinical and clinical practice.

Acknowledgements

The authors would like to thank the members in Gansu Provincial Hospital and Chinese Academy of Sciences for their technical support. This study was supported by grants from the project Youth Fund funded by Shandong Provincial Natural Science Foundation (ZR2020QC097), Project funded by China Postdoctoral Science Foundation (2023 M730723), Postdoctoral Program of Natural Science Foundation of Gansu Province (23JRRA1319), Science and technology projects of Guizhou Province (QKH-J-ZK[2021]-107, QKH-J-ZK[2022]-436), the Joint Funds of Yunnan Provincial Science and Technology Department and Kunming Medical University (202301AY070001-221), National Natural Science Foundation of China (82260031), the Natural Science Foundation of Gansu Province (21JR7RA594), Gansu Provincial Hospital Intra-Hospital Research Fund Project (22GSSYB-6), The 2022 Master/Doctor/Postdoctoral program of NHC Key Laboratory of Diagnosis and Therapy of Gastrointestinal Tumor (NHCDP2022004, NHCDP2022008), Jiangxi Provincial Novel Research & Development Institutions of Shangrao City (2020AB002, 2021F013, 2022A001, 2022AB003), Jiangxi Provincial Key New Product Incubation Program from Technical Innovation Guidance Program of Shangrao city (2020G002, 2020 K003), Spring City Plan of the High-level Talent Promotion and Training Project of Kunming (2022SCP002), and Major Science and Technology Project of Science and Technology Department of Yunnan Province (202302AA310018), Jiangxi Provincial Natural Science Foundation (20224BAB206077, 20212BAB216073), Jiangxi Provincial Leading Talent of "Double Thousand Plan" (jxsq2023102017).

Conflict of interest

The authors declare no conflict of interest.

Notes/thanks/other declarations

Not applicable.

Appendices and nomenclature

MSCs	mesenchymal stem/stromal cells
hPSCs	human pluripotent stem cells
hESCs	human embryonic stem cells
hiPSCs	human induced pluripotent stem cells
UCB-MSCs	umbilical cord blood-derived MSCs
UC-MSCs	umbilical cord-derived MSCs
ISCT	International Society for Cellular Therapy

APCs	antigen-presenting cells
TILs	tumor-infiltrating lymphocytes
IFN-γ	interferon-γ
GDEPT	gene-directed enzyme prodrug therapy
TRAMP	transgenic adenocarcinoma of the mouse prostate
CLI	critical limb ischemia
ICAM-1	intercellular adhesion molecule-1
VCAM-1	vascular cell adhesion molecule-1
SPIO	superparamagnetic iron oxide
HSV-TK	herpes simplex virus-hymidine kinase
TRAIL	TNF-related apoptosis inducing ligand
NSCLC	non-small cell lung cancer

Author details

Hao Yu[1†], Xiaonan Yang[2†], Shuang Chen[3,4†], Xianghong Xu[5], Zhihai Han[3], Hui Cai[5*], Zheng Guan[6*] and Leisheng Zhang[3,4,5,7,8,9*]

1 School of Medicine, Nankai University, Tianjin, China

2 Department of Plastic and Reconstructive Surgery and Department of Hemangioma and Vescular Malformation, Plastic Surgery Hospital Affiliated to Chinese Academy of Medical Sciences and Peiking Union Medical College, Beijing, China

3 Jiangxi Research Center of Stem Cell Engineering, Jiangxi Health-Biotech Stem Cell Technology Co., Ltd., Shangrao, China

4 Institute of Stem Cells, Health-Biotech (Tianjin) Stem Cell Research Institute Co., Ltd., Tianjin, China

5 Key Laboratory of Molecular Diagnostics and Precision Medicine for Surgical Oncology in Gansu Province and NHC Key Laboratory of Diagnosis and Therapy of Gastrointestinal Tumor, Gansu Provincial Hospital, Lanzhou, China

6 Biomedical Research Center, Affiliated Calmette Hospital of Kunming Medical University (the First Hospital of Kunming), Kunming, China

7 Center for Cellular Therapies, The First Affiliated Hospital of Shandong First Medical University, Jinan, China

8 Key Laboratory of Radiation Technology and Biophysics, Hefei Institute of Physical Science, Chinese Academy of Sciences, Hefei, China

9 Center Laboratory, The Fourth People's Hospital of Jian and The Teaching Hospital of Shandong First Medical University, Jinan, China

*Address all correspondence to: caialon@163.com; jasmin_067@163.com and leisheng_zhang@163.com

† These authors contributed equally.

References

[1] Li Y, Hermanson DL, Moriarity BS, Kaufman DS. Human iPSC-derived natural killer cells engineered with chimeric antigen receptors enhance anti-tumor activity. Cell Stem Cell. 2018;**23**(2):181-192 e185

[2] Wu H, Uchimura K, Donnelly EL, Kirita Y, Morris SA, Humphreys BD. Comparative analysis and refinement of human PSC-derived kidney organoid differentiation with single-cell Transcriptomics. Cell Stem Cell. 2018;**23**(6):869-881 e868

[3] Wu Q, Zhang L, Su P, Lei X, Liu X, Wang H, et al. MSX2 mediates entry of human pluripotent stem cells into mesendoderm by simultaneously suppressing SOX2 and activating NODAL signaling. Cell Research. 2015;**25**(12):1314-1332

[4] Zhang L, Wang H, Liu C, Wu Q, Su P, Wu D, et al. MSX2 initiates and accelerates mesenchymal stem/stromal cell specification of hPSCs by regulating TWIST1 and PRAME. Stem Cell Reports. 2018;**11**(2):497-513

[5] Wei Y, Hou H, Zhang L, Zhao N, Li C, Huo J, et al. JNKi- and DAC-programmed mesenchymal stem/stromal cells from hESCs facilitate hematopoiesis and alleviate hind limb ischemia. Stem Cell Research & Therapy. 2019;**10**(1):186

[6] Zhang L, Wei Y, Chi Y, Liu D, Yang S, Han Z, et al. Two-step generation of mesenchymal stem/stromal cells from human pluripotent stem cells with reinforced efficacy upon osteoarthritis rabbits by HA hydrogel. Cell & Bioscience. 2021;**11**(1):6

[7] Jiang B, Yan L, Wang X, Li E, Murphy K, Vaccaro K, et al. Concise review: Mesenchymal stem cells derived from human pluripotent cells, an unlimited and quality-controllable source for therapeutic applications. Stem Cells. 2019;**37**(5):572-581

[8] Yu H, Feng Y, Du W, Zhao M, Jia H, Wei Z, et al. Off-the-shelf GMP-grade UC-MSCs as therapeutic drugs for the amelioration of CCl4-induced acute-on-chronic liver failure in NOD-SCID mice. International Immunopharmacology. 2022;**113**(Pt A):109408

[9] Zhang Y, Li Y, Li W, Cai J, Yue M, Jiang L, et al. Therapeutic effect of human umbilical cord mesenchymal stem cells at various passages on acute liver failure in rats. Stem Cells International. 2018;**2018**:7159465

[10] Zhao Q, Han Z, Wang J, Han Z. Development and investigational new drug application of mesenchymal stem/stromal cells products in China. Stem Cells Translational Medicine. 2021;**10**(Suppl 2):S18-S30

[11] Zhao K, Liu Q. The clinical application of mesenchymal stromal cells in hematopoietic stem cell transplantation. Journal of Hematology & Oncology. 2016;**9**(1):46

[12] Zhao Q, Zhang L, Wei Y, Yu H, Zou L, Huo J, et al. Systematic comparison of hUC-MSCs at various passages reveals the variations of signatures and therapeutic effect on acute graft-versus-host disease. Stem Cell Research & Therapy. 2019;**10**(1):354

[13] Ankrum JA, Ong JF, Karp JM. Mesenchymal stem cells: immune evasive, not immune privileged. Nature Biotechnology. 2014;**32**(3):252-260

[14] Wu HH, Zhou Y, Tabata Y, Gao JQ. Mesenchymal stem cell-based

drug delivery strategy: From cells to biomimetic. Journal of Controlled Release. 2019;**294**:102-113

[15] Zhang X, Yang Y, Zhang L, Lu Y, Zhang Q, Fan D, et al. Mesenchymal stromal cells as vehicles of tetravalent bispecific Tandab (CD3/CD19) for the treatment of B cell lymphoma combined with IDO pathway inhibitor D-1-methyl-tryptophan. Journal of Hematology & Oncology. 2017;**10**(1):56

[16] Wang A, Zhang L, Zhao M, Yu H. Quality control and therapeutic investigations of mesenchymal stem/stromal cells during investigational new drug application for GvHD administration in China. Current Stem Cell Research & Therapy. 2023;**18**(8):1032-1040

[17] Wang M, Yuan Q, Xie L. Mesenchymal stem cell-based immunomodulation: Properties and clinical application. Stem Cells International. 2018;**2018**:3057624

[18] Yao J, Chen N, Wang X, Zhang L, Huo J, Chi Y, et al. Human supernumerary teeth-derived apical papillary stem cells possess preferable characteristics and efficacy on hepatic fibrosis in mice. Stem Cells International. 2020;**2020**:6489396

[19] Wang L, Zhang L, Liang X, Zou J, Liu N, Liu T, et al. Adipose tissue-derived stem cells from type 2 diabetics reveal conservative alterations in multidimensional characteristics. International Journal of Stem Cells. 2020;**13**(2):268-278

[20] Scioli MG, Artuso S, D'Angelo C, Porru M, D'Amico F, Bielli A, et al. Adipose-derived stem cell-mediated paclitaxel delivery inhibits breast cancer growth. PLoS One. 2018;**13**(9):e0203426

[21] Du W, Li X, Chi Y, Ma F, Li Z, Yang S, et al. VCAM-1+ placenta chorionic villi-derived mesenchymal stem cells display potent pro-angiogenic activity. Stem Cell Research & Therapy. 2016;**7**:49

[22] Wei Y, Zhang L, Chi Y, Ren X, Gao Y, Song B, et al. High-efficient generation of VCAM-1(+) mesenchymal stem cells with multidimensional superiorities in signatures and efficacy on aplastic anaemia mice. Cell Proliferation. 2020;**53**(8):e12862

[23] He Y, Guo X, Lan T, Xia J, Wang J, Li B, et al. Human umbilical cord-derived mesenchymal stem cells improve the function of liver in rats with acute-on-chronic liver failure via downregulating notch and Stat1/Stat3 signaling. Stem Cell Research & Therapy. 2021;**12**(1):396

[24] Hou H, Zhang L, Duan L, Liu Y, Han Z, Li Z, et al. Spatio-temporal Metabolokinetics and efficacy of human placenta-derived mesenchymal stem/stromal cells on mice with refractory Crohn's-like Enterocutaneous fistula. Stem Cell Reviews and Reports. 2020;**16**(6):1292-1304

[25] Yan L, Zheng D, Xu RH. Critical role of tumor necrosis factor signaling in mesenchymal stem cell-based therapy for autoimmune and inflammatory diseases. Frontiers in Immunology. 2018;**9**:1658

[26] Li E, Zhang Z, Jiang B, Yan L, Park JW, Xu RH. Generation of mesenchymal stem cells from human embryonic stem cells in a complete serum-free condition. International Journal of Biological Sciences. 2018;**14**(13):1901-1909

[27] Wang X, Lazorchak AS, Song L, Li E, Zhang Z, Jiang B, et al. Immune modulatory mesenchymal stem cells derived from human embryonic stem cells through a trophoblast-like stage. Stem Cells. 2016;**34**(2):380-391

[28] Prabha S, Merali C, Sehgal D, Nicolas E, Bhaskar N, Flores M, et al. Incorporation of paclitaxel in mesenchymal stem cells using nanoengineering upregulates antioxidant response, CXCR4 expression and enhances tumor homing. Materials Today Bio. 2023;**19**:100567

[29] Abas BI, Demirbolat GM, Cevik O. Wharton jelly-derived mesenchymal stem cell exosomes induce apoptosis and suppress EMT signaling in cervical cancer cells as an effective drug carrier system of paclitaxel. PLoS One. 2022;**17**(9):e0274607

[30] Wang XL, Zhao WZ, Fan JZ, Jia LC, Lu YN, Zeng LH, et al. Tumor tropic delivery of hyaluronic acid-poly (D,L-lactide-co-glycolide) polymeric micelles using mesenchymal stem cells for glioma therapy. Molecules. 8 Apr 2022;**27**(8):2419

[31] Borghese C, Casagrande N, Corona G, Aldinucci D. Adipose-derived stem cells primed with paclitaxel inhibit ovarian cancer spheroid growth and overcome paclitaxel resistance. Pharmaceutics. 27 Apr 2020;**12**(5):401

[32] Coccè V, Bonomi A, Cavicchini L, Sisto F, Giannì A, Farronato G, et al. Paclitaxel priming of TRAIL expressing mesenchymal stromal cells (MSCs-TRAIL) increases antitumor efficacy of their Secretome. Current Cancer Drug Targets. 15 Nov 2020. DOI: 10.2174/1568009620666201116112153 [Epub ahead of print]

[33] Coccè V, Franzè S, Brini AT, Giannì AB, Pascucci L, Ciusani E, et al. In vitro anticancer activity of extracellular vesicles (EVs) secreted by gingival mesenchymal stromal cells primed with paclitaxel. Pharmaceutics. 1 Feb 2019;**11**(2):61

[34] Kalimuthu S, Gangadaran P, Rajendran RL, Zhu L, Oh JM, Lee HW, et al. A new approach for loading anticancer drugs into mesenchymal stem cell-derived exosome Mimetics for cancer therapy. Frontiers in Pharmacology. 2018;**9**:1116

[35] Ebrahimian M, Shahgordi S, Yazdian-Robati R, Etemad L, Hashemi M, Salmasi Z. Targeted delivery of galbanic acid to colon cancer cells by PLGA nanoparticles incorporated into human mesenchymal stem cells. Avicenna Journal of Phytomedicine. 2022;**12**(3):295-308

[36] Wang J, Li M, Jin L, Guo P, Zhang Z, Zhanghuang C, et al. Exosome mimetics derived from bone marrow mesenchymal stem cells deliver doxorubicin to osteosarcoma in vitro and in vivo. Drug Delivery. 2022;**29**(1):3291-3303

[37] Wei H, Chen J, Wang S, Fu F, Zhu X, Wu C, et al. A Nanodrug consisting of doxorubicin and exosome derived from mesenchymal stem cells for osteosarcoma treatment In vitro. International Journal of Nanomedicine. 2019;**14**:8603-8610

[38] Li M, Wang J, Guo P, Jin L, Tan X, Zhang Z, et al. Exosome mimetics derived from bone marrow mesenchymal stem cells ablate neuroblastoma tumor in vitro and in vivo. Biomaterials Advances. 2022;**142**:213161

[39] Yang C, Guan Z, Pang X, Tan Z, Yang X, Li X, et al. Desialylated mesenchymal stem cells-derived extracellular vesicles loaded with doxorubicin for targeted inhibition of hepatocellular carcinoma. Cells. 25 Aug 2022;**11**(17):2642

[40] Wei H, Chen F, Chen J, Lin H, Wang S, Wang Y, et al. Mesenchymal stem cell derived exosomes as Nanodrug carrier of doxorubicin for targeted osteosarcoma therapy via SDF1-CXCR4 Axis. International Journal of Nanomedicine. 2022;**17**:3483-3495

[41] Patel N, Kommineni N, Surapaneni SK, Kalvala A, Yaun X,

Gebeyehu A, et al. Cannabidiol loaded extracellular vesicles sensitize triple-negative breast cancer to doxorubicin in both in-vitro and in vivo models. International Journal of Pharmaceutics. 2021;**607**:120943

[42] Li YS, Wu HH, Jiang XC, Zhang TY, Zhou Y, Huang LL, et al. Active stealth and self-positioning biomimetic vehicles achieved effective antitumor therapy. Journal of Controlled Release. 2021;**335**:515-526

[43] Niu G, Yousefi B, Qujeq D, Marjani A, Asadi J, Wang Z, et al. Melatonin and doxorubicin co-delivered via a functionalized graphene-dendrimeric system enhances apoptosis of osteosarcoma cells. Materials Science & Engineering. C, Materials for Biological Applications. 2021;**119**:111554

[44] Bagheri E, Abnous K, Farzad SA, Taghdisi SM, Ramezani M, Alibolandi M. Targeted doxorubicin-loaded mesenchymal stem cells-derived exosomes as a versatile platform for fighting against colorectal cancer. Life Sciences. 2020;**261**:118369

[45] Liu Y, Zhao J, Jiang J, Chen F, Fang X. Doxorubicin delivered using nanoparticles camouflaged with mesenchymal stem cell membranes to treat colon cancer. International Journal of Nanomedicine. 2020;**15**:2873-2884

[46] Gomari H, Forouzandeh Moghadam M, Soleimani M, Ghavami M, Khodashenas S. Targeted delivery of doxorubicin to HER2 positive tumor models. International Journal of Nanomedicine. 2019;**14**:5679-5690

[47] Kalimuthu S, Zhu L, Oh JM, Gangadaran P, Lee HW, Baek SH, et al. Migration of mesenchymal stem cells to tumor xenograft models and in vitro drug delivery by doxorubicin. International Journal of Medical Sciences. 2018;**15**(10):1051-1061

[48] Pessina A, Cocce V, Pascucci L, Bonomi A, Cavicchini L, Sisto F, et al. Mesenchymal stromal cells primed with paclitaxel attract and kill leukaemia cells, inhibit angiogenesis and improve survival of leukaemia-bearing mice. British Journal of Haematology. 2013;**160**(6):766-778

[49] Petrella F, Cocce V, Masia C, Milani M, Sale EO, Alessandri G, et al. Paclitaxel-releasing mesenchymal stromal cells inhibit in vitro proliferation of human mesothelioma cells. Biomedicine & Pharmacotherapy. 2017;**87**:755-758

[50] Paris JL, de la Torre P, Victoria Cabanas M, Manzano M, Grau M, Flores AI, et al. Vectorization of ultrasound-responsive nanoparticles in placental mesenchymal stem cells for cancer therapy. Nanoscale. 2017;**9**(17):5528-5537

[51] Azimifar MA, Hashemi M, Babaei N, Salmasi Z, Doosti A. Interleukin gene delivery for cancer gene therapy: In vitro and in vivo studies. Iranian Journal of Basic Medical Sciences. 2023;**26**(2):128-136

[52] Chen X, Wang K, Chen S, Chen Y. Effects of mesenchymal stem cells harboring the interferon-beta gene on A549 lung cancer in nude mice. Pathology, Research and Practice. 2019;**215**(3):586-593

[53] Du L, Liang Q, Ge S, Yang C, Yang P. The growth inhibitory effect of human gingiva-derived mesenchymal stromal cells expressing interferon-beta on tongue squamous cell carcinoma cells and xenograft model. Stem Cell Research & Therapy. 2019;**10**(1):224

[54] Du J, Liu A, Zhu R, Zhou C, Su H, Xie G, et al. The different effects of

IFN-beta and IFN-gamma on the tumor-suppressive activity of human amniotic fluid-derived mesenchymal stem cells. Stem Cells International. 2019;**2019**:4592701

[55] Byun CS, Hwang S, Woo SH, Kim MY, Lee JS, Lee JI, et al. Adipose tissue-derived mesenchymal stem cells suppress growth of Huh7 hepatocellular carcinoma cells via interferon (IFN)-beta-mediated JAK/STAT1 pathway in vitro. International Journal of Medical Sciences. 2020;**17**(5):609-619

[56] Quiroz-Reyes AG, González-Villarreal CA, Martínez-Rodriguez H, Said-Fernández S, Salinas-Carmona MC, Limón-Flores AY, et al. A combined antitumor strategy of separately transduced mesenchymal stem cells with soluble TRAIL and IFNβ produces a synergistic activity in the reduction of lymphoma and mice survival enlargement. Molecular Medicine Reports. Jun 2022;**25**(6):206

[57] Jung PY, Ryu H, Rhee KJ, Hwang S, Lee CG, Gwon SY, et al. Adipose tissue-derived mesenchymal stem cells cultured at high density express IFN-beta and TRAIL and suppress the growth of H460 human lung cancer cells. Cancer Letters. 2019;**440-441**:202-210

[58] Yenilmez EN, Genc D, Farooqi AA, Tunoglu S, Zeybek U, Akkoc T, et al. Mesenchymal stem cells combined with IFNgamma induce apoptosis of breast cancer cells partially through TRAIL. Anticancer Research. 2020;**40**(10):5641-5647

[59] Chulpanova DS, Gilazieva ZE, Kletukhina SK, Aimaletdinov AM, Garanina EE, James V, et al. Cytochalasin B-induced membrane vesicles from human mesenchymal stem cells overexpressing IL2 are able to stimulate CD8(+) T-killers to kill human triple negative breast cancer cells. Biology (Basel). 10 Feb 2021;**10**(2):141

[60] Chulpanova DS, Solovyeva VV, James V, Arkhipova SS, Gomzikova MO, Garanina EE, et al. Human mesenchymal stem cells overexpressing interleukin 2 can suppress proliferation of neuroblastoma cells in co-culture and activate mononuclear cells in vitro. Bioengineering (Basel). 17 Jun 2020;**7**(2):59

[61] Zhao C, Pu Y, Zhang H, Hu X, Zhang R, He S, et al. IL10-modified human mesenchymal stem cells inhibit pancreatic cancer growth through angiogenesis inhibition. Journal of Cancer. 2020;**11**(18):5345-5352

[62] Wu J, Xie S, Li H, Zhang Y, Yue J, Yan C, et al. Antitumor effect of IL-12 gene-modified bone marrow mesenchymal stem cells combined with Fuzheng Yiliu decoction in an in vivo glioma nude mouse model. Journal of Translational Medicine. 2021;**19**(1):143

[63] Quiroz-Reyes AG, Gonzalez-Villarreal CA, Limon-Flores AY, Delgado-Gonzalez P, Martinez-Rodriguez HG, Said-Fernandez SL, et al. Mesenchymal stem cells genetically modified by lentivirus-express soluble TRAIL and Interleukin-12 inhibit growth and reduced metastasis-relate changes in lymphoma mice model. Biomedicines. 17 Feb 2023;**11**(2):595

[64] Kulach N, Pilny E, Cichon T, Czapla J, Jarosz-Biej M, Rusin M, et al. Mesenchymal stromal cells as carriers of IL-12 reduce primary and metastatic tumors of murine melanoma. Scientific Reports. 2021;**11**(1):18335

[65] Gonzalez-Junca A, Liu FD, Nagaraja AS, Mullenix A, Lee CT, Gordley RM, et al. SENTI-101, a preparation of mesenchymal stromal

cells engineered to express IL12 and IL21, induces localized and durable antitumor immunity in preclinical models of peritoneal solid tumors. Molecular Cancer Therapeutics. 2021;**20**(9):1508-1520

[66] Wang P, Zhang J, Zhang Q, Liu F. Mesenchymal stem cells loaded with Ad5-Ki67/IL-15 enhance oncolytic adenovirotherapy in experimental glioblastoma. Biomedicine & Pharmacotherapy. 2023;**157**:114035

[67] Sun L, Wang Q, Chen B, Zhao Y, Shen B, Wang X, et al. Human gastric cancer mesenchymal stem cell-derived IL15 contributes to tumor cell epithelial-mesenchymal transition via upregulation Tregs ratio and PD-1 expression in CD4(+)T cell. Stem Cells and Development. 2018;**27**(17):1203-1214

[68] Liu X, Hu J, Li Y, Cao W, Wang Y, Ma Z, et al. Mesenchymal stem cells expressing interleukin-18 inhibit breast cancer in a mouse model. Oncology Letters. 2018;**15**(5):6265-6274

[69] Ahn J, Lee H, Seo K, Kang S, Ra J, Youn H. Anti-tumor effect of adipose tissue derived-mesenchymal stem cells expressing interferon-beta and treatment with cisplatin in a xenograft mouse model for canine melanoma. PLoS One. 2013;**8**(9):e74897

[70] Ling X, Marini F, Konopleva M, Schober W, Shi Y, Burks J, et al. Mesenchymal stem cells overexpressing IFN beta inhibit breast cancer growth and metastases through Stat3 signaling in a syngeneic tumor model. Cancer Microenvironment. 2010;**3**(1):83-95

[71] Kidd S, Caldwell L, Dietrich M, Samudio I, Spaeth EL, Watson K, et al. Mesenchymal stromal cells alone or expressing interferon-beta suppress pancreatic tumors in vivo, an effect countered by anti-inflammatory treatment. Cytotherapy. 2010;**12**(5):615-625

[72] Jin D, Jiang Y, Chang L, Wei J, Sun J. New therapeutic strategies based on biasing IL-2 mutants for cancers and autoimmune diseases. International Immunopharmacology. 2022;**110**:108935

[73] Mizui M. Natural and modified IL-2 for the treatment of cancer and autoimmune diseases. Clinical Immunology. 2019;**206**:63-70

[74] Bae J, Liu L, Moore C, Hsu E, Zhang A, Ren Z, et al. IL-2 delivery by engineered mesenchymal stem cells re-invigorates CD8(+) T cells to overcome immunotherapy resistance in cancer. Nature Cell Biology. 2022;**24**(12):1754-1765

[75] Gao W, Pan J, Pan J. Antitumor activities of Interleukin-12 in melanoma. Cancers (Basel). 14 Nov 2022;**14**(22):5592

[76] Ryu CH, Park SH, Park SA, Kim SM, Lim JY, Jeong CH, et al. Gene therapy of intracranial glioma using interleukin 12-secreting human umbilical cord blood-derived mesenchymal stem cells. Human Gene Therapy. 2011;**22**(6):733-743

[77] Zhou Y, Husman T, Cen X, Tsao T, Brown J, Bajpai A, et al. Interleukin 15 in cell-based cancer immunotherapy. International Journal of Molecular Sciences. 30 Jun 2022;**23**(13):7311

[78] Jing W, Chen Y, Lu L, Hu X, Shao C, Zhang Y, et al. Human umbilical cord blood-derived mesenchymal stem cells producing IL15 eradicate established pancreatic tumor in syngeneic mice. Molecular Cancer Therapeutics. 2014;**13**(8):2127-2137

[79] Li Z, Yu X, Werner J, Bazhin AV, D'Haese JG. The role of interleukin-18

in pancreatitis and pancreatic cancer. Cytokine & Growth Factor Reviews. 2019;**50**:1-12

[80] Kuppala MB, Syed SB, Bandaru S, Varre S, Akka J, Mundulru HP. Immunotherapeutic approach for better management of cancer—Role of IL-18. Asian Pacific Journal of Cancer Prevention. 2012;**13**(11):5353-5361

[81] Kim N, Nam YS, Im KI, Lim JY, Lee ES, Jeon YW, et al. IL-21-expressing mesenchymal stem cells prevent lethal B-cell lymphoma through efficient delivery of IL-21, which redirects the immune system to target the tumor. Stem Cells and Development. 2015;**24**(23):2808-2821

[82] Klimova D, Jakubechova J, Altanerova U, Nicodemou A, Styk J, Szemes T, et al. Extracellular vesicles derived from dental mesenchymal stem/stromal cells with gemcitabine as a cargo have an inhibitory effect on the growth of pancreatic carcinoma cell lines in vitro. Molecular and Cellular Probes. 2023;**67**:101894

[83] Tibensky M, Jakubechova J, Altanerova U, Pastorakova A, Rychly B, Baciak L, et al. Gene-directed enzyme/prodrug therapy of rat brain tumor mediated by human mesenchymal stem cell suicide gene extracellular vesicles in vitro and in vivo. Cancers (Basel). 31 Jan 2022;**14**(3):735

[84] Blanco-Fernandez B, Cano-Torres I, Garrido C, Rubi-Sans G, Sanchez-Cid L, Guerra-Rebollo M, et al. Engineered microtissues for the bystander therapy against cancer. Materials Science & Engineering. C, Materials for Biological Applications. 2021;**121**:111854

[85] Kenarkoohi A, Bamdad T, Soleimani M, Soleimanjahi H, Fallah A, Falahi S. HSV-TK expressing mesenchymal stem cells exert inhibitory effect on cervical cancer model. International Journal of Molecular and Cellular Medicine. 2020;**9**(2):146-154

[86] Pastorakova A, Jakubechova J, Altanerova U, Altaner C. Suicide gene therapy mediated with exosomes produced by mesenchymal stem/stromal cells stably transduced with HSV thymidine kinase. Cancers (Basel). 28 Apr 2020;**12**(5):1096

[87] Duhrsen L, Hartfuss S, Hirsch D, Geiger S, Maire CL, Sedlacik J, et al. Preclinical analysis of human mesenchymal stem cells: Tumor tropism and therapeutic efficiency of local HSV-TK suicide gene therapy in glioblastoma. Oncotarget. 2019;**10**(58):6049-6061

[88] Wei D, Hou J, Zheng K, Jin X, Xie Q, Cheng L, et al. Suicide gene therapy against malignant gliomas by the local delivery of genetically engineered umbilical cord mesenchymal stem cells as cellular vehicles. Current Gene Therapy. 2019;**19**(5):330-341

[89] Yang J, Lv K, Sun J, Guan J. Anti-tumor effects of engineered mesenchymal stem cells in colon cancer model. Cancer Management and Research. 2019;**11**:8443-8450

[90] Malik YS, Sheikh MA, Xing Z, Guo Z, Zhu X, Tian H, et al. Polylysine-modified polyethylenimine polymer can generate genetically engineered mesenchymal stem cells for combinational suicidal gene therapy in glioblastoma. Acta Biomaterialia. 2018;**80**:144-153

[91] Malekshah OM, Sarkar S, Nomani A, Patel N, Javidian P, Goedken M, et al. Bioengineered adipose-derived stem cells for targeted enzyme-prodrug therapy of ovarian cancer intraperitoneal metastasis. Journal of Controlled Release. 2019;**311-312**:273-287

[92] Krasikova LS, Karshieva SS, Cheglakov IB, Belyavsky AV. Mesenchymal stem cells expressing cytosine deaminase inhibit growth of murine melanoma B16F10 in vivo. Molekuliarnaia Biologiia (Mosk). 2015;**49**(6):1007-1015

[93] NguyenThai QA, Sharma N, Luong do H, Sodhi SS, Kim JH, Kim N, et al. Targeted inhibition of osteosarcoma tumor growth by bone marrow-derived mesenchymal stem cells expressing cytosine deaminase/5-fluorocytosine in tumor-bearing mice. The Journal of Gene Medicine. 2015;**17**(3-5):87-99

[94] Krassikova LS, Karshieva SS, Cheglakov IB, Belyavsky AV. Combined treatment, based on lysomustine administration with mesenchymal stem cells expressing cytosine deaminase therapy, leads to pronounced murine Lewis lung carcinoma growth inhibition. The Journal of Gene Medicine. 2016;**18**(9):220-233

[95] Nouri FS, Wang X, Hatefi A. Genetically engineered theranostic mesenchymal stem cells for the evaluation of the anticancer efficacy of enzyme/prodrug systems. Journal of Controlled Release. 2015;**200**:179-187

[96] Chung T, Na J, Kim YI, Chang DY, Kim YI, Kim H, et al. Dihydropyrimidine dehydrogenase is a prognostic marker for mesenchymal stem cell-mediated cytosine deaminase gene and 5-Fluorocytosine prodrug therapy for the treatment of recurrent gliomas. Theranostics. 2016;**6**(10):1477-1490

[97] Abrate A, Buono R, Canu T, Esposito A, Del Maschio A, Luciano R, et al. Mesenchymal stem cells expressing therapeutic genes induce autochthonous prostate tumour regression. European Journal of Cancer. 2014;**50**(14):2478-2488

[98] Chang DY, Jung JH, Kim AA, Marasini S, Lee YJ, Paek SH, et al. Combined effects of mesenchymal stem cells carrying cytosine deaminase gene with 5-fluorocytosine and temozolomide in orthotopic glioma model. American Journal of Cancer Research. 2020;**10**(5):1429-1441

[99] Ullah M, Kuroda Y, Bartosh TJ, Liu F, Zhao Q, Gregory C, et al. Erratum: iPS-derived MSCs from an expandable bank to deliver a prodrug-converting enzyme that limits growth and metastases of human breast cancers. Cell Death Discovery. 2017;**3**:17029

[100] Oishi T, Ito M, Koizumi S, Horikawa M, Yamamoto T, Yamagishi S, et al. Efficacy of HSV-TK/GCV system suicide gene therapy using SHED expressing modified HSV-TK against lung cancer brain metastases. Molecular Therapy—Methods & Clinical Development. 2022;**26**:253-265

[101] Metz MZ, Gutova M, Lacey SF, Abramyants Y, Vo T, Gilchrist M, et al. Neural stem cell-mediated delivery of irinotecan-activating carboxylesterases to glioma: Implications for clinical use. Stem Cells Translational Medicine. 2013;**2**(12):983-992

[102] Choi SA, Lee JY, Wang KC, Phi JH, Song SH, Song J, et al. Human adipose tissue-derived mesenchymal stem cells: Characteristics and therapeutic potential as cellular vehicles for prodrug gene therapy against brainstem gliomas. European Journal of Cancer. 2012;**48**(1):129-137

[103] Alizadeh Zeinabad H, Szegezdi E. TRAIL in the treatment of cancer: From soluble cytokine to nanosystems. Cancers (Basel). 19 Oct 2022;**14**(20):5125

[104] Voss OH, Arango D, Tossey JC, Villalona Calero MA, Doseff AI. Splicing reprogramming of TRAIL/DISC-components sensitizes lung cancer cells

to TRAIL-mediated apoptosis. Cell Death & Disease. 2021;**12**(4):287

[105] Chulpanova DS, Pukhalskaia TV, Gilazieva ZE, Filina YV, Mansurova MN, Rizvanov AA, et al. Cytochalasin B-induced membrane vesicles from TRAIL-overexpressing mesenchymal stem cells induce extrinsic pathway of apoptosis in breast cancer mouse model. Current Issues in Molecular Biology. 2023;**45**(1):571-592

[106] Chen F, Zhong X, Dai Q, Li K, Zhang W, Wang J, et al. Human umbilical cord MSC delivered-soluble TRAIL inhibits the proliferation and promotes apoptosis of B-ALL cell in vitro and in vivo. Pharmaceuticals (Basel). 11 Nov 2022;**15**(11):1391

[107] Sun L, Wang J, Wang Q, He Z, Sun T, Yao Y, et al. Pretreatment of umbilical cord derived MSCs with IFN-gamma and TNF-alpha enhances the tumor-suppressive effect on acute myeloid leukemia. Biochemical Pharmacology. 2022;**199**:115007

[108] Un Choi Y, Yoon Y, Jung PY, Hwang S, Hong JE, Kim WS, et al. TRAIL-overexpressing adipose tissue-derived mesenchymal stem cells efficiently inhibit tumor growth in an H460 xenograft model. Cancer Genomics Proteomics. 2021;**18**(4):569-578

[109] Park SA, Han HR, Ahn S, Ryu CH, Jeun SS. Combination treatment with VPA and MSCs-TRAIL could increase anti-tumor effects against intracranial glioma. Oncology Reports. 2021;**45**(3):869-878

[110] Liu Z, Li S, Ma T, Zeng J, Zhou X, Li H, et al. Secreted TRAIL gene-modified adipose-derived stem cells exhibited potent tumor-suppressive effect in hepatocellular carcinoma cells. Immunity, Inflammation and Disease. 2021;**9**(1):144-156

[111] Deng L, Wang C, He C, Chen L. Bone mesenchymal stem cells derived extracellular vesicles promote TRAIL-related apoptosis of hepatocellular carcinoma cells via the delivery of microRNA-20a-3p. Cancer Biomarkers. 2021;**30**(2):223-235

[112] Salmasi Z, Hashemi M, Mahdipour E, Nourani H, Abnous K, Ramezani M. Mesenchymal stem cells engineered by modified polyethylenimine polymer for targeted cancer gene therapy, in vitro and in vivo. Biotechnology Progress. 2020;**36**(6):e3025

[113] Zakaria N, Yahaya BH. Adipose-derived mesenchymal stem cells promote growth and migration of lung adenocarcinoma cancer cells. Advances in Experimental Medicine and Biology. 2020;**1292**:83-95

[114] Xue T, Wang X, Ru J, Zhang L, Yin H. The inhibitory effect of human umbilical cord mesenchymal stem cells expressing anti-HAAH scFv-sTRAIL fusion protein on glioma. Frontiers in Bioengineering and Biotechnology. 2022;**10**:997799

[115] Choi SA, Lee C, Kwak PA, Park CK, Wang KC, Phi JH, et al. Histone deacetylase inhibitor panobinostat potentiates the anti-cancer effects of mesenchymal stem cell-based sTRAIL gene therapy against malignant glioma. Cancer Letters. 2019;**442**:161-169

[116] Wang Z, Chen H, Wang P, Zhou M, Li G, Hu Z, et al. Site-specific integration of TRAIL in iPSC-derived mesenchymal stem cells for targeted cancer therapy. Stem Cells Translational Medicine. 2022;**11**(3):297-309

[117] Niess H, von Einem JC, Thomas MN, Michl M, Angele MK, Huss R, et al. Treatment of advanced gastrointestinal

tumors with genetically modified autologous mesenchymal stromal cells (TREAT-ME1): Study protocol of a phase I/II clinical trial. BMC Cancer. 2015;**15**:237

[118] Garcia-Castro J, Alemany R, Cascallo M, Martinez-Quintanilla J, Arriero Mdel M, Lassaletta A, et al. Treatment of metastatic neuroblastoma with systemic oncolytic virotherapy delivered by autologous mesenchymal stem cells: An exploratory study. Cancer Gene Therapy. 2010;**17**(7):476-483

Chapter 4

Hematopoietic Development of Human Pluripotent Stem Cells

Igor M. Samokhvalov and Anna Liakhovitskaia

Abstract

Blood development proceeds through several waves of hematopoietic progenitors with unclear lineage relationships, which convolute the understanding of the process. Thinking of the hematopoietic precursors as the "blood germ layer" can integrate these waves into a unified hematopoietic lineage that originates in the yolk sac, the earliest site of blood development. Hematopoietic differentiation of pluripotent stem cells (PSCs) reflects to a certain extent the complexities of the yolk sac hematopoiesis. In the unified version of blood issue development, the PSC-derived hematopoiesis can also generate post-yolk sac hematopoietic progenitors. To do this, the differentiation has to be arranged for the reproduction of the intraembryonic hematopoiesis. Inflammatory signaling was recently shown to be actively engaged in blood ontogenesis. In addition, a highly recapitulative differentiation of human PSCs was found to spontaneously ignite intense sterile inflammation that has both instructive and destructive roles in the hPSC-hematopoiesis. Inflammatory induction of blood progenitors during hPSC-derived hematopoietic development has to be properly contained. A possible explanation of problems associated with *in vitro* blood development is the failure of inflammation containment and resolution.

Keywords: pluripotent stem cells, hematopoietic differentiation, inflammation, hematopoietic stem cells, tissue macrophages

1. Introduction

Hematopoietic differentiation initiated by human pluripotent stem cells (hPSCs) is a surrogate model for studying early human hematopoietic development. It is generally accepted that the transition of PSCs into blood cells essentially recapitulates the yolk sac stage of hematopoiesis [1, 2], which turns out to be significantly more complicated than previously thought. Most of the information on blood origin came from mouse studies despite the fact that mouse development is highly specialized and adapted to specific living conditions of the species. The mouse hematopoietic system emerges in the early ontogenesis to support embryo development after the initiation of the heartbeat. The first blood cells arise directly from extraembryonic mesoderm at murine embryonic day 7.0 (E7.0) and consist of primitive erythrocytes or erythroblasts, primitive megakaryocytes, and macrophages that perform the tissue oxygenation, preventing embryonic blood loss, and clearing of apoptotic cells, respectively [3]. The role of progenitors for these cell types is taken by mesodermal precursors of the yolk sac.

A second wave of erythromyeloid progenitors (EMPs) initiates in the yolk sac at E8.0–8.5, and progenitors with lymphoid or lymphomyeloid potential emerge in the yolk sac and embryo proper at around E9.5 [3–7]. These data picture an obvious trend of yolk sac hematopoiesis from primitive monopotential and bipotential progenitors [8–10] through EMPs toward the lymphoid-primed multipotent progenitors (LMPPs). This fast evolution of hematopoietic potentials does not culminate in the self-renewing multipotential progenitors, although it does not mean that precursors of these progenitors do not emerge in the yolk sac. Hematopoietic stem cells (HSCs) that can self-renew and repopulate the hematopoietic system of a conditioned recipient are first observed at E10.5, emerging in the aorta–gonad–mesonephros (AGM) region [11]. At this stage, very few HSCs arise alongside numerous HSC-independent hematopoietic progenitors from the endothelium of the dorsal aorta in an extremely peculiar process designated as endothelial to hematopoietic transition (EHT) [12, 13]. EHT is thought to be also involved in the formation of EMPs and possibly LMPPs in the yolk sac [3, 14]. Circulating HSCs and their precursors colonize the extravascular territories of the fetal liver (FL) at E12.0–12.5, where they undergo massive expansion before migrating to incipient bone marrow at E17.5, the main site of hematopoiesis during the adulthood [15, 16]. Of note, the FL colonization looks very similar to leukocyte extravasation during inflammation [17], and the FL niche harbors and cultivates both HSC-dependent as well as HSC-independent progenitors during development.

Human hematopoiesis development is far less amenable for systemic molecular and cellular studies but, based on several lines of evidence, follows a similar logic. Primordial angioblastic cords appear in the human yolk sac around Day 16 of gestation [18] whereas the formation of blood islands, observed starting from Day 19 [19], leads to the production of primitive erythrocytes and macrophages [20], as well as primitive megakaryocytes [21]. Clonogenic multilineage hematopoietic progenitors, the analogs of murine EMPs, are functionally identified in the human yolk sac at four to five weeks of gestation [22, 23] when systemic circulation already started. Under the influence of the discovery that the AGM region is a niche of the first mouse HSCs, early human studies had also presumed the intraembryonic origin of adult hematopoiesis and HSCs [24], although the autonomous generation of HSCs in the human AGM region was not confirmed [25].

Human PSC-hematopoiesis struggles to generate authentic HSCs [26] perhaps due to the yolk sac mode of the differentiation. The yolk sac-like concept of hPSC-hematopoiesis, however, does not overrule the possibility of HSC generation *in vitro*. There are two main reasons for the continuation of the efforts: (1) no solid data prove that the yolk sac does not produce precursors of HSCs, while the opposite may be correct [27]; (2) only primary hPSC differentiation is similar to the yolk sac hematopoiesis, and additional steps with selected cell populations may improve the HSC chances.

This chapter will discuss the role of archetypal cell interactions of sterile inflammation in hPSC-hematopoiesis. A recapitulative hPSC differentiation reproduces the inflammatory mechanisms that participate in the *in vivo* hematopoietic development. The spontaneous sterile inflammation and inefficient resolution can be important factors contributing to the difficulties in the derivation of HSCs from hPSCs.

2. Induction of hematopoiesis in the human conceptus

Early human development is drastically and conceptually different from the development of the mouse embryo, the standard developmental model. In mice, a number of fate mapping and molecular studies have established that during the blastocyst formation ICM cells lose their potential for the trophectoderm specification. In contrast, more ontogenically advanced epiblast cells in human preimplantation embryos still showed outstanding developmental plasticity. Pre-gastrulation human epiblast was demonstrated to give rise not only to all three germ layers but also to the trophectoderm [28]. These findings provided compelling experimental evidence of the regulative embryonic development in humans and showed that the regulative development, inherent to all mammals and birds, is more prominently manifested in humans compared to mice. The increased developmental plasticity may be an adaptation to the long and relatively slow gestation so that a majority of developmental deviations or failures can be effectively fixed by other embryonic cells. Alternatively, stricter lineage bifurcation in the mouse embryo suits better to fast and short-term ontogenesis. The outstanding human plasticity may not be limited to epiblast cells. Later conceptal locations, including extraembryonic mesoderm with the emerging hematopoietic system, are likely to possess the increased regulative potential, although experimental proof of the postimplantation plasticity is difficult to obtain.

Due to the limited access to the tissues of the early human conceptus, the initiation and structure of the human hematopoietic development is less well-known compared to that of the mouse. In mice, the founding members of the blood germ layer are located in the pre-gastrulation epiblast giving rise to the primitive blood and hemogenic endothelial cells (HECs) within the yolk sac vascular endothelium [29]. A body of evidence [18, 30–32] indicates that human hypoblast or primitive endoderm rather than epiblast gives rise to the extraembryonic mesoderm (EXM) of the primary yolk sac. The anatomical origin of EXM had been debated and other anatomical regions of the pre-gastrulation embryo were thought to contribute to its emergence [18, 33]. Nevertheless, the most compelling cell tracing evidence points to the hypoblast origin [32]. Close alignment of EXM and epiblast transcriptomes in monkey pre-streak embryo [34] is an indication of the early commitment of epiblast cells to mesodermal specification before gastrulation. It is also possible that the secondary mesoderm arising from the primitive streak during gastrulation also contributes to expanding EXM [35], although the extent of this contribution is unknown. The major role of EXM is the generation of first blood cells and vasculature in primordial structures traditionally called blood islands. These tissues sustain the developing human embryo for a period, but this is not the only function of the EXM derivatives. Mouse studies demonstrate the long-term contribution of the yolk sac hematopoiesis into various cohorts of adult cells [36–38], including adult HSCs [27, 39].

The hypoblast origin of the early hematopoiesis implies that the human blood tissue develops in parallel with primitive streak-derived secondary mesoderm. The origin of the yolk sac blood islands is polyclonal [40], otherwise, there would be not enough cells to sustain embryonic development by providing erythroblast-based tissue oxygenation and timely clearing of cell debris by primitive macrophages. This polyclonal, definitive mesoderm-independent development means that blood induction is a morphogenic event analogous to the formation of ectoderm, trophectoderm, and gastrulation. Then, EXM and its progeny—blood can be considered a cryptic or atavistic germ layer containing all elements required for its development and function.

The concept of the blood germ layer helps to alleviate an obvious problem of the strange mode of hematopoietic ontogenesis with two seemingly separate developmental lineages, the primitive and the definitive hematopoiesis. In the blood germ layer hypothesis, EXM is a founder conceptus location for the whole of the hematopoietic lineage developing as a special germ layer. As in the classical germ layers, blood has transitory primordial tissue, the primitive blood, within primordial organs, blood islands, that contain precursors of definitive tissue. These precursors represented by HECs use the vascular niche for their survival and maintenance of the undifferentiated state to serve as an origin of definitive tissues and organs. During ontogenesis, the derivatives of the definitive HECs migrate throughout the vascular system probing the potential niches for settling down and being sometimes or somewhere retained on vascular beds due to the affinity of their surface molecules to cognate receptors expressed on endothelial cells. Later on, the blood germ layer, in collaboration with the derivatives of other germ layers, develops into the immune system with its multiple organs and hematopoietic system settled in the bone marrow. In mice, under strong selective pressure to minimize the gestation period, the regulative development canceled the primary mesoderm and fused the blood layer to the secondary mesoderm.

Supporting the blood germ layer concept, recent studies show that the yolk sac hematopoiesis is more complicated than was previously thought. Early evidence of the key role of the yolk sac in the initiation of fetal and adult hematopoiesis [41–43] was dismissed based on experimental data supporting the intraembryonic origin of definitive hematopoiesis, which was located in the pre-circulation para-aortic splanchnopleura and its derivative—the AGM region. The presence of definitive EMPs and LMPPs in the yolk sac was rediscovered by the new post-AGM generation of researchers using modern approaches and techniques. The lineage tracing experiments in mice demonstrated that populations of tissue-resident macrophages in the adult, including Kupffer cells, alveolar macrophages, and microglia develop from the progenitors originating in the yolk sac [36–38, 44]. The role of the yolk sac in the establishment of adult cell populations became even more apparent after it was found that subpopulations of mast and T cells descend from hematopoietic programs localized in the yolk sac [45–47]. The unifying feature of the abovementioned adult, yolk sac-derived cell populations is their capacity to maintain themselves within the tissues by lifelong self-renewal. Taken together, these data indicate that at least some yolk sac progenitors can acquire self-renewal potential during development. In other words, the ability to self-renew is not something impossible for yolk sac cells to gain on their way toward adulthood. It is then not too untrustworthy that cell tracing and gene reactivation studies indicate the yolk sac origin of the pre-HSCs [27, 39].

The signaling events participating in the induction of human blood are still poorly understood. In the mESC differentiation model, Wnt/β-catenin signaling promotes, whereas Notch signaling suppresses primitive erythroblast development [48]. It was also found that the transient expression of Numb in mesodermal precursors led to the inhibition of the Notch signaling. BMP4 promotes the generation of VEGFR2$^+$ cells within the mESC-derived lateral mesoderm and VEGF supports the subsequent specification and expansion of hematopoietic and endothelial cells [49–51]. Short-term exposure to BMP4 was also instrumental in driving the mesodermal specification of human ESCs [52]. It is highly likely that BMP4 is among the human conceptal factors that induce EXM and help to initiate the hematopoiesis transition. Recent hPSC differentiation data showed that FGF2, BMP4, VEGF, and possibly signals evoked by the collagen IV adherence are sufficient to induce efficient primitive and definitive human hematopoiesis [53].

3. Hematopoietic differentiation of hPSCs

Since their derivation in 1998, human embryonic stem cells (hESCs) have become a useful model for studying human development and molecular mechanisms of lineage specification *in vitro*. Reprogramming human somatic cells into embryonic-like stem cells [54], named human induced pluripotent stem cells (hiPSCs), opened realistic perspectives for generating various therapeutic cell populations, disease modeling, and drug discovery [55]. The hope is boosted by experiments, in which mouse iPSC-derived primitive macrophages differentiated into tissue-resident macrophages and microglia upon injection into recipient animals [56]. Nevertheless, common usage of instructive hematopoietic cytokines shifts the ontogenic program of PSC-derived mesodermal precursors into the development of cells with a limited functional diversity [57].

Conventional, not naïve, hPSCs that possess so-called primed pluripotency are considered to represent postimplantation epiblast. In order to begin their way to hematopoiesis these cells have to transit from epiblast-like though hypoblast-like to mesodermal epigenetic state under the influence of internal and external effector molecules. This transition is facilitated by a standard variety of factors, and over the decades, there were no significant changes in basic approaches to the hematopoietic differentiation of hPSCs. The traditional methods include coculture with supportive stromal cells, most prominently the M-CSF-negative OP9 cell line [58, 59]; various modifications of the planar differentiation [60, 61]; 3D cultures through the formation of embryoid bodies (EBs) [62, 63], or a combination of 2D and 3D differentiation [53, 64]. In the well-substantiated trend, the protocols are modified in order to remove the fetal calf serum from the culture medium. The *in vitro* analog of the primary mesoderm is generally induced through the BMP signaling [52, 65, 66], followed by hemogenic endothelium induction by VEGF, SCF, FLT3L, IL3, and FGF2 [67], or similar recipes of exogenous growth factors and hematopoietic cytokines. The shift toward hematopoiesis is first manifested by the expression of CD235a, CD43, and CD34, a marker of hematopoietic progenitors and vascular endothelial cells. The endothelium is always present in the hematopoietic differentiation of PSCs since both lineages have common ancestor cells: the hemangioblasts—a mouse primitive streak mesendodermal entity [68], and hemogenic endothelial cells, VEGFR2-positive mesodermal precursors expressing CD34 and the endothelial marker VE-cadherin [69, 70]. CD34 is expressed at a substantially higher level on endothelial cells compared to emerging $CD43^+$ hematopoietic progenitors. The objective of the first step of differentiation is usually to obtain $CD34^+CD45^+$ cells that can be used as hematopoietic progenitors in downstream cultures or applications. Omitting the hematopoietic cytokines and minimizing the use of the mesodermal/endothelial growth factors can strongly improve the recovery of the clonogenic hematopoietic progenitors, including the multilineage ones [53, 64]. The hPSC-derived progenitors can be used to generate clinically relevant blood and immune cell populations [71, 72]. Nonetheless, a practical biotechnological design for getting large numbers of safe functional cells is missing. There is also a serious concern about the genomic stability of ethically acceptable induced pluripotent stem cells, which threaten to prevent the efficient use of hPSC differentiation for therapeutical purposes [73].

The search for a powerful inducer of hematopoiesis led to the conclusion that BMP4 signaling plays a key role in the induction of blood-competent mesoderm upon hPSC differentiation. In mice, there is compelling evidence that Bmp4 signaling is critically required for mesodermal induction [50, 74–79]. In the recently developed model of the early human peri-implantation development, BMP4 was shown to participate in the maintenance of EXM [80]. Induction of EXM was achieved by inhibition of Nodal signaling and GSK3β. This suppression only indirectly reflects the induction events in the developing human embryo and actual signals leading to the suppression are unknown. We, therefore, can only guess which factors induce EXM in the peri-implantation embryo. It seems that the BMP4 signaling does not only participate in the maintenance of EXM *in vitro* but also may induce the emergence of the tissue. Indeed, previous studies showed that Nodal inhibition results in enhanced BMP4 signaling in the pluripotent stem cell context [81]. Activin/Nodal/TGF-β pathway was not active in the naïve hPSC-derived EXM cells, and no data were available on the role of the FGF2 signaling.

In addition to BMP4, hematopoietic induction in hPSC-derived extraembryonic mesoderm is strongly dependent on VEGF, a growth factor of outstanding pleiotropy. In human development, VEGF treatment enhances blastocyst outgrowth and stimulates embryo implantation [82]. In postnatal development, accumulating evidence indicates the crucial role of VEGF in body growth and organ development [83]. Other findings demonstrate that VEGF also promotes neurogenesis, neuronal patterning, neuroprotection, and glial growth independently of the angiogenic function of the growth factor [84]. The major developmental function of VEGF is organizing and stimulating embryonic vasculogenesis and angiogenesis [85, 86]. In inflammation, VEGF and one of its receptors, VEGFR1, previously identified as a decoy receptor, were found to play a role in the recruitment and activation of monocytes and macrophages [87, 88]. In hPSC biology, VEGF signaling participates in the mesendodermal induction of hESCs [89] and selectively promotes erythropoietic development from hESCs, which have been strongly augmented by BMP4 [90]. Together with FGF2, VEGF is crucial for the progression of mesoderm to hemogenic endothelium [91], which was identified as $CD31^{+}CD34^{+}VE\text{-}CADHERIN^{+}KDR^{+}$ cell population [92]. These hPSC-HECs can initiate both primitive and broadly defined definitive hematopoiesis.

Strictly defined definitive hematopoiesis arises only from fetal or adult-type hematopoietic stem cells (HSCs). Generation of lymphoid cells from PSCs was previously considered proof of the definitive lineage potential, although most likely it recapitulates the emergence of pre-HSC hematopoiesis of the yolk sac type. The derivation of HSCs of adult or fetal type remains elusive despite many efforts, including the most recent one [93]. One possible reason for the HSC failure is a low recapitulative quality of hPSC hematopoietic differentiation *in vitro*. HSC precursors change several locations within developing conceptus until they land in the bone marrow [94]. During their development, the HSC lineage cells seem to focus on segregation from progenitor-driven embryonic hematopoiesis and settling in a safe haven of fetal liver, spleen, and bone marrow sinusoids. Therefore, a good recapitulation of HSC development should include a second leg of differentiation so that pre-HSCs are isolated from active hPSC-hematopoiesis. The secondary differentiation most likely should include a stromal culture supported by selected cytokines and small molecules such as stable derivatives of ascorbic acid.

Excessive use of growth factors and cytokines at a nonphysiological concentration to induce the emergence of EXMCs and HECs is another reason for the poor generation of multipotent hematopoietic progenitors and HSCs by differentiating hPSCs. We do not know the exact makeup of the signaling factors participating in the hematopoietic induction within the developing human conceptus. Moreover, inductive events in early mammalian embryos are performed by short-range signaling proteins and growth factors [65]. Therefore, any use of exogenous hematopoietic cytokines to initiate hPSC-derived hematopoiesis may *a priori* disturb the developmental pathway of HSC precursors due to the strong instructive influence of these cytokines [95]. The published protocols use complex compositions and excessive concentrations of cytokines, growth factors, and signaling molecules to ensure efficient conversion of hPSCs into hematopoietic cells at the cost of a proper recapitulation of hematopoietic development in the conceptus. Such an approach decreases the value of the hPSC differentiation as a model of early hematopoietic ontogenesis. Furthermore, the use of external cytokines accelerates terminal differentiation of hematopoietic progenitors that negatively influences the length of the proliferative stage and the yield of clinically relevant cellular material.

In the structural aspect of hPSC-hematopoiesis, planar, 3D, or stromal differentiations do not reproduce to any acceptable extent the anatomy of the peri-implantation human embryo. In the early postimplantation human embryo, EXM cells spread from the 3D embryo proper part over the distal trophoblast in an essentially planar fashion (**Figure 1A**). Therefore, the optimal recapitulation of mesodermal induction

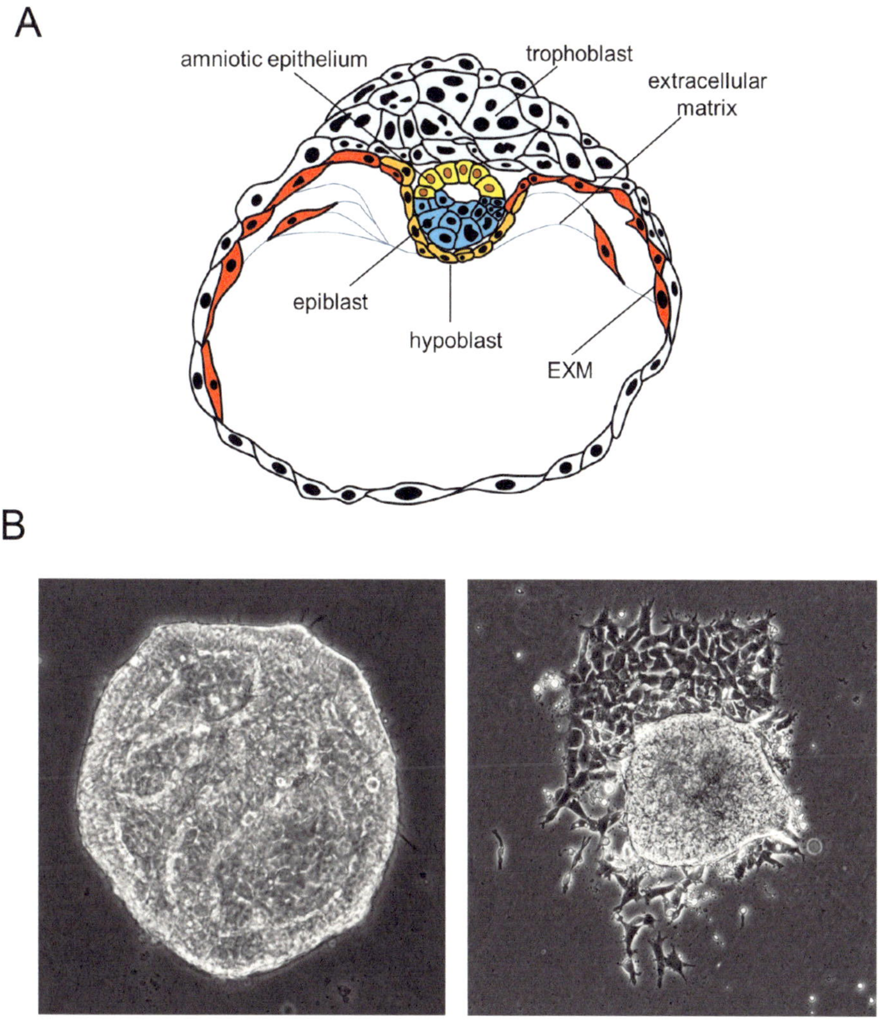

Figure 1.
Forced attachment of hPSC-EBs reproduces the early EXM development. (A) Day 8–9 human peri-implantation embryo. The induction of EXM. (B) Early stages of hPSC differentiation. Shortly after attachment to the collagen surface, EB undergoes epithelialization (left panel). Next day, the EB-derived EXM starts to spread over the surface.

in peri-implantation embryos should include the spreading of differentiating hPSCs from a central 3D cell mass in a planar circumferential fashion. Stromal support from adult sources, the fetal liver, and the AGM region creates artificial differentiation conditions that are not encountered at the start of hematopoietic development.

In the recently published protocol for recapitulative differentiation of hPSCs, no external cytokines were used [53]. The onset of EXM development was reproduced by forced attachment of hPSC-EBs to collagen-covered surfaces in the presence of BMP4, VEGF, and mTeSR1-derived FGF2. The attachment initiated active planar spreading of mesenchymal cells (**Figure 1B**) followed by spontaneous formation of blood island-like aggregates (**Figure 2**) in which hematopoietic induction occurs. Differentiating cultures produced a variety of endogenous cytokines that supported hematopoietic development and the production of large numbers of progenitors.

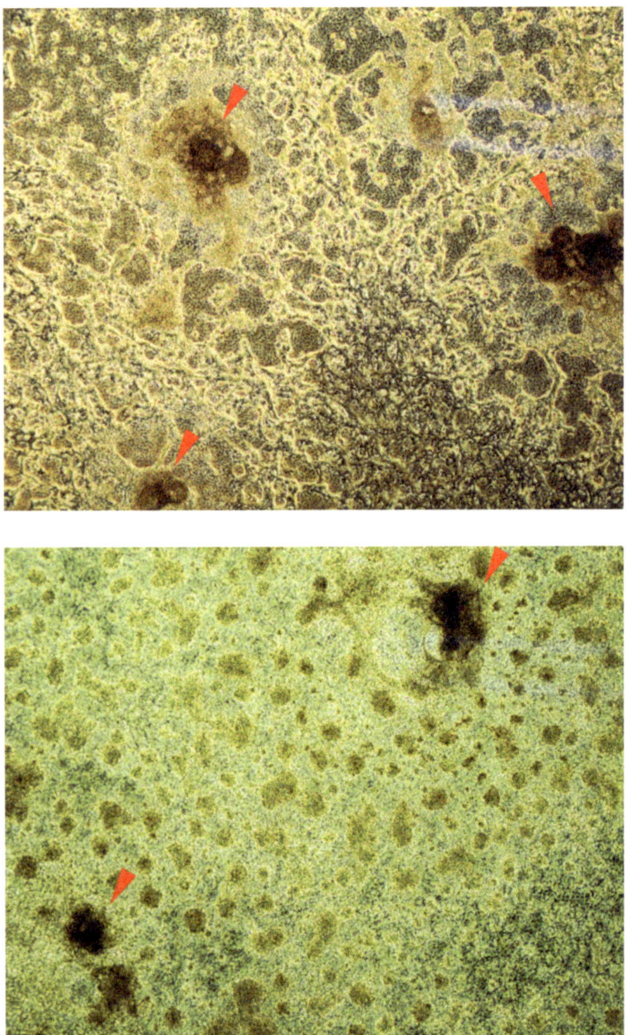

Figure 2.
After attachment to collagen-coated surface, hPSC-derived embryonic bodies form angioblastic cords (upper panel) that transform later on into blood island-like structures (lower panel).

Among others, M-CSF, a pro-inflammatory cytokine [96], was secreted at an exceptionally high level. After 1 week of the differentiation, emerging myeloid cells activate genetic programs that induce and control spontaneous sterile inflammation.

4. Inflammation

Inflammation is a complex, local and systemic, defensive, and adaptive response of the immune system triggered by a variety of agents that disturb homeostasis on an organismal or cellular level. The inflammatory factors include pathogens, damaged cells, toxic compounds, irradiation, and irritation. The variety of the factors boils down to the intrusion of external organisms and the noninfectious damage to tissue cells. Cell damage and infectious agents activate inflammatory cells and trigger inflammatory signaling pathways. In most cases, these are the NF-κB, MAPK, and JAK-STAT signaling [97]. The cellular challenges are sensed by special sentinel receptor molecules that activate the immune system.

Damage-associated molecular patterns (DAMPs) such as oxidized lipoproteins, HMGB1, S100 calcium-binding proteins, heat-shock proteins, or pathogen-associated molecular patterns (PAMPs) such as uncapped viral RNA, lipopeptides, flagellin, and lipopolysaccharides are identified by pattern recognition receptors (PRRs) on the cell surface or endosomes in the cytoplasm. These PRRs are represented by a vast variety of cell surface and cytoplasmic molecules, including the TLRs (Toll-like receptors), the CLRs (C-type lectin receptors), NLRs (NOD-like receptors), the RLRs (RIG1-like receptors), and the RAGE (receptor for advanced glycosylation endproducts). PRRs are preferentially expressed on sentinel immune cells, including mast cells, macrophages, dendritic cells, innate lymphoid cells, and basophils. In addition, many tissues have nonimmune-tissue sentinel cells [98, 99]. Binding the DAMPs/PAMPs ligands to cognate PRRs activates the NF-κB transcriptional effector complex, which then induces the expression of inflammatory cytokines. These cytokines then initiate the cellular phase of the inflammatory reaction. The key step is upregulating cell adhesion molecules on endothelial cells. VCAM-1, ICAM-1, and E-selectin, all inducible by inflammatory cytokines, promote, in cooperation with chemokines and other endothelial adhesion molecules, the adherence and extravasation, or diapedesis, of neutrophils and monocytes into damaged tissue. In sterile inflammation, the major post-diapedesis role belongs to monocytes, which differentiate into tissue macrophages that clean up cellular and extracellular matrix debris.

To preserve tissue homeostasis, acute inflammation has to be suppressed to avoid persistent, chronic inflammation that leads to additional and broader tissue damage. Inflammatory neutrophils are major culprits in collateral tissue and cell breakdown [100]. Stable chemokine gradients may attract an excess of neutrophils even when the microbial infection is already contained. These granulocytes discharge their immense cytotoxic arsenal into the extracellular space of surrounding tissues even if they fail to encounter a microbial agent for a short period of time. Similar cell damage occurs upon neutrophil activation and degranulation in sterile inflammation conditions [101]. Neutrophils release their own set of proteases, activate proteases that are expressed in a latent form by cells resident in the tissues, and inactivate anti-proteases by oxidation [102, 103] using reactive oxygen species (ROS). In addition to neutrophils, activated macrophages initiate nitric oxide-dependent killing of resident cells [104].

Inflammation resolution is an active, well-coordinated process that normally initiates shortly after the start of an inflammatory reaction. It involves the spatially- and temporally-controlled production of specialized pro-resolving mediators (SPMs) [105], which coincide with a gradual dilution of chemokine gradients across an inflamed tissue. In consequence, neutrophil recruitment becomes attenuated and eventually stops, and then programmed death by apoptosis is engaged. Apoptotic neutrophils are cleared by macrophage phagocytosis followed by the release of anti-inflammatory and reparative cytokines such as IL-10 and TGFβ1, which can suppress pro-inflammatory signaling from Toll-like receptors [106, 107]. The inflammation resolution sequence ends with the conversion of macrophages into the M2 reparative type and/or the departure of macrophages through the lymphatic vessels. Phagocytosis of apoptotic cells inhibits activated macrophage killing of resident tissue cells and triggers the secretion of VEGF, which participates in the repair of endothelial and epithelial injury.

5. Inflammatory hematopoietic development

Most of the research on the role of inflammatory signaling in hematopoiesis is concentrated on the emergence and regulation of HSCs. Adult HSCs are capable not just to respond to inflammatory signals but also to secrete pro-inflammatory/anti-inflammatory cytokines and chemokines [108, 109], including IFN-α [110, 111], IFN-γ [112], TNF-α [113], TGF-β [114], IL-1, and IL-6 [115]. All these cytokines are pleiotropic, and their expression in the blood stem cells may not translate into classical inflammatory cell interactions. However, the expression of PRRs and their co-receptors in HSCs and hematopoietic progenitors [116] directly indicates the involvement of the progenitor domain in sensing stress situations in the hematopoietic system. Hematopoietic progenitors respond to the ligation of TLRs by entering cell cycling, proliferation, and differentiation, and, therefore, can be considered as a part of the innate immune system. In the developmental aspect, hematopoietic progenitors, including HSCs, may arise as highly specialized innate immune cells possessing substantial proliferation and differentiation potential. Indirectly, this notion is supported by another mouse study, which has shown that HSCs in the AGM region exhibit lower levels of IFN-α expression compared to fetal liver HSCs, and this trait can contribute to the lower engraftment potential of the AGM HSCs. Similar to innate immune cells, these HSCs strongly reacted to IFN-α treatment by improvement of their proliferation capacity upon transplantation [117].

Homeostatic expression of another interferon, IFN-γ, a powerful anti-viral cytokine, also activates the proliferation of HSCs *in vivo* during normal hematopoiesis [112]. However, in the hPSC differentiation model, exogenous IFN-γ failed to stimulate the emergence of $CD34^+$ HECs and $CD34^+CD43^+CD45^+CD235a^-$ definitive hematopoietic cells [118]. The iconic pro-inflammatory cytokine IL-1β affects murine adult hematopoiesis by accelerating HSC proliferation and myeloid differentiation through activation of the PU.1 gene program [119]. TNF-α has been implied to play an important role in hematopoietic development based on its abundant expression in the murine yolk sac and fetal liver [120]. Zebrafish studies strongly indicate TNF-α importance for HSC emergence and specification [121]. The data is especially interesting because it implies the involvement of primitive neutrophils in the maintenance and emergence of HSCs in the zebrafish AGM region. Nevertheless, IL-1β and TNF-α signaling did not improve the hematopoietic differentiation of hPSCs [118]. The most straightforward explanation of IFN-γ, IL-1β, and TNF-α failure to influence hPSC-derived hematopoiesis is a non-recapitulative differentiation protocol, although, in the more positive attitude, it is possible that strong and instructive inflammatory signaling was spontaneously activated during the hematopoietic transition of hPSCs and additional external stimulation could not have any visible effect.

The notion of spontaneous sterile inflammation is supported by recent bioinformatics studies of highly recapitulative hematopoietic differentiation of hPSCs [53]. In the study, efficient hematopoiesis was induced and supported by hematopoietic cytokines produced endogenously in differentiated cultures. Many of the secreted cytokines, such as IL-8, IL-11, IL-16, and M-CSF, were pro-inflammatory, some of them were detected before the emergence of hematopoietic cells. These cytokines programmed the early hematopoietic cells to create an inflammatory milieu that included neutrophil activation, T cell activation, response to bacterial molecular patterns, activation and regulation of innate immune response, phagocytosis, viral response, and others. These observations strongly suggest the instructive role of inflammation in the recapitulative hPSC-hematopoiesis. Such hematopoiesis, however, cannot be sustained long-term in the primary hPSC differentiation probably due to the failure of inflammation resolution, so poorly controlled neutrophil activation leads to the gradual destruction of generated hematopoietic progenitors and blood cells. The optimal solution is to recapitulate the post-yolk sac phase of embryonic hematopoiesis by transferring early hPSC-derived hematopoietic progenitors into secondary differentiating cultures supported by stromal cells. The transfer allows the progenitors to escape from the inflammatory environment before the neutrophil activation. In the secondary differentiation, the escapees undergo extensive proliferation and can develop into lymphoid cells and HSCs in proper culture conditions.

The induction of sterile inflammatory programs in the hPSC differentiation is mediated by DAMPs released from dead or dying hPSCs. Many hPSC cannot easily enter the commitment sequence and have to undergo apoptosis. The major effector of the sterile inflammation is possibly BMP4, the growth factor that is required for the initiation of hematopoietic development. Corroborating evidence [122–124] shows that BMP4 signaling is linked to the induction of inflammatory nuclear factor-κB (NF-κB), nicotinamide adenine dinucleotide phosphate oxidase-1 (NOX1), and intracellular adhesions molecule-1 (ICAM-1), which are the key factors of inflammation initiation. This inflammatory role of BMP4 was described in the adult context, but it demonstrated that there exist molecular mechanisms involving BMP4 in the inflammatory response. It is unknown whether BMP4 induces inflammation at early stages of embryo development, but it is safer to minimize both the concentration and duration of the BMP4 treatment to avoid its participation in the ignition of a potent inflammatory reaction, including neutrophil activation and degranulation.

6. Conclusions

Accumulating evidence suggests that hematopoiesis develops as a single germ layer. Emerging hematopoietic progenitor cells adapt to the constantly changing conceptal environment by modulating their proliferative and self-renewal potential. They have to use inflammatory mechanisms to penetrate endothelial barriers and settle into transitory or permanent niches. Human PSC-derived hematopoiesis, despite evident problems, has the capacity to develop the self-renewal potential but has to reproduce the *in vivo* development as close as possible.

Inflammatory signaling can play an instructive role in hPSC-derived hematopoiesis. It is not just an artifact of the culture and may reflect some aspects of the hematopoietic development in the conceptus. Inflammation can also negatively influence the expansion of the hematopoietic populations due to the failure of regulated resolution. For successful cell engineering, the early hematopoietic progenitors have to be removed from the primary differentiation culture and used in the secondary differentiation with the immunosuppressive environment.

Acknowledgement

This work was supported by the Russian Science Foundation, grant #22-15-20063.

Conflict of interest

The authors declare no conflict of interest.

Author details

Igor M. Samokhvalov[1]* and Anna Liakhovitskaia[2]

1 Engineering Center of Genetic and Cellular Biotechnology, V.I. Vernadsky Crimean Federal University, Simferopol, Crimea

2 Faculty of Health Sciences, Bristol Medical School, Bristol, UK

*Address all correspondence to: igormikhail@netscape.net

References

[1] Kennedy M, D'Souza SL, Lynch-Kattman M, Schwantz S, Keller G. Development of the hemangioblast defines the onset of hematopoiesis in human ES cell differentiation cultures. Blood. 2007;**109**:2679-2687. DOI: 10.1182/blood-2006-09-047704

[2] Zambidis ET, Peault B, Park TS, Bunz F, Civin CI. Hematopoietic differentiation of human embryonic stem cells progresses through sequential hematoendothelial, primitive, and definitive stages resembling human yolk sac development. Blood. 2005;**106**:860-870. DOI: 10.1182/blood-2004-11-4522

[3] Palis J. Hematopoietic stem cell-independent hematopoiesis: Emergence of erythroid, megakaryocyte, and myeloid potential in the mammalian embryo. FEBS Letters. 2016;**590**:3965-3974. DOI: 10.1002/1873-3468.12459

[4] Yoshimoto M, Montecino-Rodriguez E, Ferkowicz MJ, Porayette P, Shelley WC, Conway SJ, et al. Embryonic day 9 yolk sac and intra-embryonic hemogenic endothelium independently generate a B-1 and marginal zone progenitor lacking B-2 potential. Proceedings of the National Academy of Sciences of the USA. 2011;**108**:1468-1473. DOI: 10.1073/pnas.1015841108

[5] Yoshimoto M, Porayette P, Glosson NL, Conway SJ, Carlesso N, Cardoso AA, et al. Autonomous murine T-cell progenitor production in the extra-embryonic yolk sac before HSC emergence. Blood. 2012;**119**:5706-5714. DOI: 10.1182/blood-2011-12-397489

[6] Boiers C, Carrelha J, Lutteropp M, Luc S, Green JC, Azzoni E, et al. Lymphomyeloid contribution of an immune-restricted progenitor emerging prior to definitive hematopoietic stem cells. Cell Stem Cell. 2013;**13**:535-548. DOI: 10.1016/j.stem.2013.08.012

[7] Ghosn E, Yoshimoto M, Nakauchi H, Weissman IL, Herzenberg LA. Hematopoietic stem cell-independent hematopoiesis and the origins of innate-like B lymphocytes. Development. 2019;**146**:dev170571. DOI: 10.1242/dev.170571

[8] Palis J, Robertson S, Kennedy M, Wall C, Keller G. Development of erythroid and myeloid progenitors in the yolk sac and embryo proper of the mouse. Development. 1999;**126**:5073-5081. DOI: 10.1242/dev.126.22.5073

[9] Tober J, Koniski A, McGrath KE, Vemishetti R, Emerson R, de Mesy-Bentley KK, et al. The megakaryocyte lineage originates from hemangioblast precursors and is an integral component both of primitive and of definitive hematopoiesis. Blood. 2007;**109**:1433-1441. DOI: 10.1182/blood-2006-06-031898

[10] Tober J, McGrath KE, Palis J. Primitive erythropoiesis and megakaryopoiesis in the yolk sac are independent of c-myb. Blood. 2008;**111**:2636-2639. DOI: 10.1182/blood-2007-11-124685

[11] Medvinsky A, Dzierzak E. Definitive hematopoiesis is autonomously initiated by the AGM region. Cell. 1996;**86**:897-906. DOI: 10.1016/s0092-8674(00)80165-8

[12] Ottersbach K. Endothelial-to-haematopoietic transition: An update on the process of making blood. Biochemical Society Transactions. 2019;**47**:591-601. DOI: 10.1042/BST20180320

[13] Zhu Q, Gao P, Tober J, Bennett L, Chen C, Uzun Y, et al. Developmental trajectory of prehematopoietic stem cell formation from endothelium. Blood. 2020;**136**:845-856. DOI: 10.1182/blood.2020004801

[14] Frame JM, Fegan KH, Conway SJ, McGrath KE, Palis J. Definitive hematopoiesis in the yolk sac emerges from Wnt-responsive hemogenic endothelium independently of circulation and arterial identity. Stem Cells. 2016;**34**:431-444. DOI: 10.1002/stem.2213

[15] Ema H, Nakauchi H. Expansion of hematopoietic stem cells in the developing liver of a mouse embryo. Blood. 2000;**95**:2284-2288. DOI: 10.1182/blood.V95.7.2284

[16] Gao X, Xu C, Asada N, Frenette PS. The hematopoietic stem cell niche: From embryo to adult. Development. 2018;**145**:dev139691. DOI: 10.1242/dev.139691

[17] Samokhvalov IM. Deconvoluting the ontogeny of hematopoietic stem cells. Cellular and Molecular Life Sciences. 2014;**71**:957-978. DOI: 10.1007/s00018-013-1364-7

[18] Luckett WP. Origin and differentiation of the yolk sac and extraembryonic mesoderm in presomite human and rhesus monkey embryos. American Journal of Anatomy. 1978;**152**:59-97. DOI: 10.1002/aja.1001520106

[19] Tavian M, Hallais MF, Peault B. Emergence of intraembryonic hematopoietic precursors in the pre-liver human embryo. Development. 1999;**126**:793-803. DOI: 10.1242/dev.126.4.793

[20] Bloom W, Bartelmez GW. Hematopoiesis in young human embryos. American Journal of Anatomy. 1940;**67**:21-53. DOI: 10.1002/aja.1000670103

[21] Fukuda T. Fetal hemopoiesis. I. Electron microscopic studies on human yolk sac hemopoiesis. Virchows Archiv B: Cell Pathology. 1973;**14**:197-213

[22] Migliaccio G, Migliaccio AR, Petti S, Mavilio F, Russo G, Lazzaro D, et al. Human embryonic hemopoiesis. Kinetics of progenitors and precursors underlying the yolk sac----liver transition. Journal of Clinical Investigation. 1986;**78**:51-60. DOI: 10.1172/JCI112572

[23] Huyhn A, Dommergues M, Izac B, Croisille L, Katz A, Vainchenker W, et al. Characterization of hematopoietic progenitors from human yolk sacs and embryos. Blood. 1995;**86**:4474-4485. DOI: 10.1182/blood.V86.12.4474.bloodjournal86124474

[24] Tavian M, Robin C, Coulombel L, Peault B. The human embryo, but not its yolk sac, generates lympho-myeloid stem cells: Mapping multipotent hematopoietic cell fate in intraembryonic mesoderm. Immunity. 2001;**15**:487-495. DOI: 10.1016/s1074-7613(01)00193-5

[25] Ivanovs A, Rybtsov S, Welch L, Anderson RA, Turner ML, Medvinsky A. Highly potent human hematopoietic stem cells first emerge in the intraembryonic aorta-gonad-mesonephros region. Journal of Experimental Medicine. 2011;**208**:2417-2427. DOI: 10.1084/jem.20111688

[26] Ditadi A, Sturgeon CM, Keller G. A view of human haematopoietic development from the Petri dish. Nature Reviews Molecular Cell Biology. 2017;**18**:56-67. DOI: 10.1038/nrm.2016.127

[27] Tanaka Y, Hayashi M, Kubota Y, Nagai H, Sheng G, Nishikawa S-I,

et al. Early ontogenic origin of the hematopoietic stem cell lineage. Proceedings of the National Academy of Sciences of the USA. 2012;**109**:4515-4520. DOI: 10.1073/pnas.1115828109

[28] Guo G, Stirparo GG, Strawbridge SE, Spindlow D, Yang J, Clarke J, et al. Human naïve epiblast cell possess unrestricted lineage potential. Cell Stem Cell. 2021;**28**:1040-1056. DOI: 10.1016/j.stem.2021.02.025

[29] Padrón-Barthe L, Temiño S, Villa del Campo C, Carramolino L, Isern J, Torres M. Clonal analysis identifies hemogenic endothelium as the source of the blood-endothelial common lineage in the mouse embryo. Blood. 2014;**124**:2523-2532. DOI: 10.1182/blood-2013-12-545939

[30] Takashina T. Haemopoiesis in the human yolk sac. Journal of Anatomy. 1987;**151**:125-135

[31] Nakamura T, Okamoto I, Sasaki K, Yabuta Y, Iwatani C, Tsuchiya H, et al. A developmental coordinate of pluripotency among mice, monkeys and humans. Nature. 2016;**537**:57-62. DOI: 10.1038/nature19096

[32] Spencer Chapman M, Ranzoni AM, Myers B, Williams N, Coorens THH, Mitchell E, et al. Lineage tracing of human development through somatic mutations. Nature. 2021;**595**:85-90. DOI: 10.1038/s41586-021-03548-6

[33] Hertig AT, Rock J, Adams EC. A description of 34 human ova within the first 17 days of development. American Journal of Anatomy. 1956;**98**:435-493. DOI: 10.1002/aja.1000980306

[34] Yang R, Goedel A, Kang Y, Si C, Chu C, Zheng Y, et al. Amnion signals are essential for mesoderm formation in primates. Nature Communications. 2021;**12**:5126. DOI: 10.1038/s41467-021-25186-2

[35] Ross C, Boroviak TE. Origin and function of the yolk sac in primate embryogenesis. Nature Communications. 2020;**11**:3760. DOI: 10.1038/s41467-020-17575-w

[36] Ginhoux F, Greter M, Leboeuf M, Nandi S, See P, Gokhan S, et al. Fate mapping analysis reveals that adult microglia derive from primitive macrophages. Science. 2010;**330**:841-845. DOI: 10.1126/science.1194637

[37] Schulz C, Gomez Perdiguero E, Chorro L, Szabo-Rogers H, Cagnard N, Kierdorf K, et al. A lineage of myeloid cells independent of Myb and hematopoietic stem cells. Science. 2012;**336**:86-90. DOI: 10.1126/science.1219179

[38] Gomez Perdiguero E, Klapproth K, Schulz C, Busch K, Azzoni E, Crozet L, et al. Tissue-resident macrophages originate from yolk-sac-derived erythro-myeloid progenitors. Nature. 2015;**518**:547-551. DOI: 10.1038/nature13989

[39] Samokhvalov IM, Samokhvalova NI, Nishikawa S-I. Cell tracing shows the contribution of the yolk sac to adult haematopoiesis. Nature. 2007;**446**:1056-1061. DOI: 10.1038/nature05725

[40] Ueno H, Weissman IL. Clonal analysis of mouse development reveals a polyclonal origin for yolk sac blood islands. Developmental Cell. 2006;**11**:519-533. DOI: 10.1016/j.devcel.2006.08.001

[41] Moore MA, Metcalf D. Ontogeny of the haemopoietic system: Yolk sac origin of in vivo and in vitro colony forming cells in the developing mouse embryo. British Journal of Haematology. 1970;**18**:279-296. DOI: 10.1111/j.1365-2141.1970.tb01443.x

[42] Moore MAS, Owen JJT. Stem-cell migration in developing myeloid and lymphoid systems. Lancet. 1967;**2**:658-659. DOI: 10.1016/S0140-6736(67)90693-9

[43] Weissman I, Papaioannou V, Gardner R. Fetal hematopoietic origin of the adult hematolymphoid system. In: Clarkson B, Marks PA, Till JE, editors. Differentiation of Normal and Neoplastic Hematopoietic Cells. Cold Spring Harbor, NY: Cold Spring Harbor Laboratory Press; 1978. pp. 33-43

[44] Mass E, Ballesteros I, Farlik M, Halbritter F, Günther P, Crozet L, et al. Specification of tissue-resident macrophages during organogenesis. Science. 2016;**353**:aaf4238. DOI: 10.1126/science.aaf4238

[45] Gentek R, Ghigo C, Hoeffel G, Bulle MJ, Msallam R, Gautier G, et al. Hemogenic endothelial fate mapping reveals dual developmental origin of mast cells. Immunity. 2018;**48**:1160-1171.e5. DOI: 10.1016/j.immuni.2018.04.025

[46] Li Z, Liu S, Xu J, Zhang X, Han D, Liu J, et al. Adult connective tissue-resident mast cells originate from late erythro-myeloid progenitors. Immunity. 2018;**49**:640-653.e5. DOI: 10.1016/j.immuni.2018.09.023

[47] Gentek R, Ghigo C, Hoeffel G, Jorquera A, Msallam R, Wienert S, et al. Epidermal γδ T cells originate from yolk sac hematopoiesis and clonally self-renew in the adult. Journal of Experimental Medicine. 2018;**215**:2994-3005. DOI: 10.1084/jem.20181206

[48] Cheng X, Huber TL, Chen VC, Gadue P, Keller GM. Numb mediates the interaction between Wnt and Notch to modulate primitive erythropoietic specification from the hemangioblast. Development. 2008;**135**:3447-3458. DOI: 10.1242/dev.025916

[49] Choi K, Kennedy M, Kazarov A, Papadimitriou JC, Keller G. A common precursor for hematopoietic and endothelial cells. Development. 1998;**125**:725-732. DOI: 10.1242/dev.125.4.725

[50] Park C, Afrikanova I, Chung YS, Zhang WJ, Arentson E, Fong GG, et al. A hierarchical order of factors in the generation of FLK1- and SCL expressing hematopoietic and endothelial progenitors from embryonic stem cells. Development. 2004;**131**:2749-2762. DOI: 10.1242/dev.01130

[51] Nostro MC, Cheng X, Keller GM, Gadue P. Wnt, activin, and BMP signaling regulate distinct stages in the developmental pathway from embryonic stem cells to blood. Cell Stem Cell. 2008;**2**:60-71. DOI: 10.1016/j.stem.2007.10.011

[52] Zhang P, Li J, Tan Z, Wang C, Liu T, Chen L, et al. Short-term BMP4 treatment initiates mesoderm induction in human embryonic stem cells. Blood. 2008;**111**:1933-1941. DOI: 10.1182/blood-2007-02-074120

[53] Philonenko ES, Tan Y, Wang C, Zhang B, Shah Z, Zhang J, et al. Recapitulative haematopoietic development of human pluripotent stem cells in the absence of exogenous haematopoietic cytokines. Journal of Cellular and Molecular Medicine. 2021;**25**:8701-8714. DOI: 10.1111/jcmm.16826

[54] Takahashi K, Tanabe K, Ohnuki M, Narita M, Ichisaka T, Tomoda K, et al. Induction of pluripotent stem cells from adult human fibroblasts by defined factors. Cell. 2007;**131**:861-872. DOI: 10.1016/j.cell.2007.11.019

[55] Takahashi K, Yamanaka S. A decade of transcription factor-mediated reprogramming to pluripotency. Nature Reviews Molecular Cell Biology. 2016;**17**:183-193. DOI: 10.1038/nrm.2016.8

[56] Takata K, Kozaki T, Zhe Wei Lee C, Thion MS, Otsuka M, Lim S, et al. Induced-pluripotent-stem-cell-derived primitive macrophages provide a platform for modeling tissue-resident macrophage differentiation and function. Immunity. 2017;**47**:183-198. DOI: 10.1016/j.immuni.2017.06.017

[57] Lee CZW, Kozaki T, Ginhoux F. Studying tissue macrophages in vitro: Are iPSC-derived cells the answer? Nature Reviews Immunology. 2018;**18**:716-725. DOI: 10.1038/s41577-018-0054-y

[58] Holmes R, Zúñiga-Pflücker JC. The OP9-DL1 system: Generation of T-lymphocytes from embryonic or hematopoietic stem cells in vitro. Cold Spring Harbor Protocols. 2009;**4**:pdb.prot5156. DOI: 10.1101/pdb.prot5156

[59] Ditadi A, Sturgeon CM. Directed differentiation of definitive hemogenic endothelium and hematopoietic progenitors from human pluripotent stem cells. Methods San Diego California. 2016;**101**:65-72. DOI: 10.1016/j.ymeth.2015.10.001

[60] Salvagiotto G, Burton S, Daigh CA, Rajesh D, Slukvin II, Seay NJ. A defined, feeder-free, serum-free system to generate in vitro hematopoietic progenitors and differentiated blood cells from hESCs and hiPSCs. PLoS One. 2011;**6**:e17829. DOI: 10.1371/journal.pone.0017829

[61] Slukvin II. Hematopoietic specification from human pluripotent stem cells: Current advances and challenges toward de novo generation of hematopoietic stem cells. Blood. 2013;**122**:4035-4046. DOI: 10.1182/blood-2013-07-474825

[62] Cerdan C, Hong SH, Bhatia M. Formation and hematopoietic differentiation of human embryoid bodies by suspension and hanging drop cultures. Current Protocols in Stem Cell Biology. 2007;**Chapter 1**:Unit 1D.2. DOI: 10.1002/9780470151808.sc01d02s3

[63] Ng ES, Davis RP, Hatzistavrou T, Stanley EG, Elefanty AG. Directed differentiation of human embryonic stem cells as spin embryoid bodies and a description of the hematopoietic blast colony forming assay. Current Protocols in Stem Cell Biology. 2008;**Chapter 1**:Unit 1D.3. DOI: 10.1002/9780470151808.sc01d03s4

[64] Shah Z, Filonenko ES, Ramensky V, Fan C, Wang C, Ullah H, et al. MYB bi-allelic targeting abrogates primitive clonogenic progenitors while the emergence of primitive blood is not affected. Haematologica. 2021;**106**:2191-2202. DOI: 10.3324/haematol.2020.249193

[65] Beddington RS, Robertson EJ. Axis development and early asymmetry in mammals. Cell. 1999;**96**:195-209. DOI: 10.1016/s0092-8674(00)80560-7

[66] Langdon YG, Mullins MC. Maternal and zygotic control of zebrafish dorsoventral axial patterning. Annual Review of Genetics. 2011;**45**:357-377. DOI: 10.1146/annurev-genet-110410-132517

[67] Ackermann M, Liebhaber S, Klusmann J-H, Lachmann N. Lost in translation: Pluripotent stem cell-derived hematopoiesis. EMBO Molecular Medicine. 2015;**7**:1388-1402. DOI: 10.15252/emmm.201505301

[68] Huber TL, Kouskoff V, Fehling HJ, Palis J, Keller G. Haemangioblast commitment is initiated in the primitive streak of the mouse embryo. Nature. 2004;**432**:625-630. DOI: 10.1038/nature03122

[69] Nishikawa SI, Nishikawa S, Hirashima M, Matsuyoshi N, Kodama H. Progressive lineage analysis by cell sorting and culture identifies FLK1+VE-cadherin+ cells at a diverging point of endothelial and hemopoietic lineages. Development. 1998;**125**:1747-1757. DOI: 10.1242/dev.125.9.1747

[70] Ditadi A, Sturgeon CM, Tober J, Awong G, Kennedy M, Yzaguirre AD, et al. Human definitive haemogenic endothelium and arterial vascular endothelium represent distinct lineages. Nature Cell Biology. 2015;**17**:580-591. DOI: 10.1038/ncb3161

[71] Serra M, Brito C, Correia C, Alves PM. Process engineering of human pluripotent stem cells for clinical applications. Trends in Biotechnology. 2012;**30**:350-359. DOI: 10.1016/j.tibtech.2012.03.003

[72] Ilic D, Ogilvie C. Pluripotent stem cells in clinical setting-new development and overview of current status. Stem Cells. 2022;**40**:791-801. DOI: 10.1093/stmcls/sxac040

[73] Keller A, Spits C. The impact of acquired genetic abnormalities on the clinical translation of human pluripotent stem cells. Cell. 2021;**10**:3246. DOI: 10.3390/cells10113246

[74] Johansson BM, Wiles MV. Evidence for involvement of activin A and bone morphogenetic protein 4 in mammalian mesoderm and hematopoietic development. Molecular Cell Biology. 1995;**15**:141-151. DOI: 10.1128/MCB.15.1.141

[75] Wiles MV, Johansson BM. Analysis of factors controlling primary germ layer formation and early hematopoiesis using embryonic stem cell in vitro differentiation. Leukemia. 1997;**11**(Suppl 3):454-456 PMID: 9209423

[76] Kramer J, Hegert C, Guan K, Wobus AM, Muller PK, Rohwedel J. Embryonic stem cell-derived chondrogenic differentiation in vitro: Activation by BMP-2 and BMP-4. Mechanisms of Development. 2000;**92**:193-205. DOI: 10.1016/s0925-4773(99)00339-1

[77] Nakayama N, Lee J, Chiu L. Vascular endothelial growth factor synergistically enhances bone morphogenetic protein-4-dependent lymphohematopoietic cell generation from embryonic stem cells in vitro. Blood. 2000;**95**:2275-2283. DOI: 10.1182/blood.V95.7.2275

[78] Loebel DAF, Watson CM, De Young RA, Tam PPL. Lineage choice and differentiation in mouse embryos and embryonic stem cells. Developmental Biology. 2003;**264**:1-14. DOI: 10.1016/s0012-1606(03)00390-7

[79] Suzuki A, Raya A, Kawakami Y, Morita M, Matsui T, Nakashima K, et al. Nanog binds to Smad1 and blocks bone morphogenetic protein-induced differentiation of embryonic stem cells. Proceedings of the National Academy of Sciences of the USA. 2006;**103**:10294-10299. DOI: 10.1073/pnas.0506945103

[80] Pham TXA, Panda A, Kagawa H, To SK, Ertekin C, Georgolopoulos G, et al. Modeling human extraembryonic mesoderm cells using naïve pluripotent stem cells. Cell Stem Cell. 2022;**29**:1346-1365.e10. DOI: 10.1016/j.stem.2022.08.001

[81] Galvin KE, Travis ED, Yee D, Magnuson T, Vivian JL. Nodal signaling

regulates the bone morphogenic protein pluripotency pathway in mouse embryonic stem cells. Journal of Biological Chemistry. 2010;**285**:19747-19756. DOI: 10.1074/jbc.M109.077347

[82] Hannan NJ, Paiva P, Meehan KL, Rombauts LJ, Gardner DK, Salamonsen LA. Analysis of fertility-related soluble mediators in human uterine fluid identifies VEGF as a key regulator of embryo implantation. Endocrinology. 2011;**152**:4948-4956. DOI: 10.1210/en.2011-12

[83] Guo X, Yi H, Li TC, Wang Y, Wang H, Chen X. Role of vascular endothelial growth factor (VEGF) in human embryo implantation: Clinical implications. Biomolecules. 2021;**11**:253. DOI: 10.3390/biom11020253

[84] Rosenstein JM, Krum JM, Ruhrberg C. VEGF in the nervous system. Organogenesis. 2010;**6**:107-114. DOI: 10.4161/org.6.2.11687

[85] Carmeliet P, Ferreira V, Breier G, Pollefeyt S, Kieckens L, Gertsenstein M, et al. Abnormal blood vessel development and lethality in embryos lacking a single VEGF allele. Nature. 1996;**380**:435-439. DOI: 10.1038/380435a0

[86] Ferrara N, Carver-Moore K, Chen H, Dowd M, Lu L, O'Shea KS, et al. Heterozygous embryonic lethality induced by targeted inactivation of the VEGF gene. Nature. 1996;**380**:439-442. DOI: 10.1038/380439a0

[87] Zittermann SI, Issekutz AC. Endothelial growth factors VEGF and bFGF differentially enhance monocyte and neutrophil recruitment to inflammation. Journal of Leukocyte Biology. 2006;**80**:247-257. DOI: 10.1189/jlb.1205718

[88] Weddell JC, Chen S, Imoukhuede PI. VEGFR1 promotes cell migration and proliferation through PLCγ and PI3K pathways. NPJ System Biology and Applications. 2017;**4**:1. DOI: 10.1038/s41540-017-0037-9

[89] Xin C, Zhu C, Jin Y, Li H. Discovering the role of VEGF signaling pathway in mesendodermal induction of human embryonic stem cells. Biochemical and Biophysical Research Communications. 2021;**553**:58-64. DOI: 10.1016/j.bbrc.2021.03.036

[90] Cerdan C, Rouleau A, Bhatia M. VEGF-A_{165} augments erythropoietic development from human embryonic stem cells. Blood. 2004;**103**:2504-2512. DOI: 10.1182/blood-2003-07-2563

[91] Bruveris FF, Ng ES, Stanley EG, Elefanty AG. VEGF, FGF2, and BMP4 regulate transitions of mesoderm to endothelium and blood cells in a human model of yolk sac hematopoiesis. Experimental Hematology. 2021;**103**:30-39.e2. DOI: 10.1016/j.exphem.2021.08.006

[92] Atkins MH, Scarfò R, McGrath KE, Yang D, Palis J, Ditadi A, et al. Modeling human yolk sac hematopoiesis with pluripotent stem cells. Journal of Experimental Medicine. 2021;**219**:e20211924. DOI: 10.1084/jem.20211924

[93] Zhu Y, Wang T, Gu J, Huang K, Zhang T, Zhang Z, et al. Characterization and generation of human definitive multipotent hematopoietic stem/progenitor cells. Cell Discovery. 2020;**6**:89. DOI: 10.1038/s41421-020-00213-6

[94] Mikkola HKA, Orkin SH. The journey of developing hematopoietic stem cells. Development. 2006;**133**:3733-3744. DOI: 10.1242/dev.02568

[95] Rieger MA, Hoppe PS, Smejkal BM, Eitelhuber AC, Schroeder T.

Hematopoietic cytokines can instruct lineage choice. Science. 2009;**325**:217-218. DOI: 10.1126/science.1171461

[96] Kremlev SG, Chapoval AI, Evans R. CSF-1 (M-CSF) enhances the inflammatory response of fibronectin-primed macrophages: Pathways involved in activation of the cytokine network. Nature Immunology. 1998;**16**:228-243. DOI: 10.1159/000069449

[97] Chen L, Deng H, Cui H, Fang J, Zuo Z, Deng J, et al. Inflammatory responses and inflammation-associated diseases in organs. Oncotarget. 2018;**9**:7204-7218. DOI: 10.18632/oncotarget.23208

[98] Andonegui G, Zhou H, Bullard D, Kelly MM, Mullaly SC, McDonald B, et al. Mice that exclusively express TLR4 on endothelial cells can efficiently clear a lethal systemic Gram-negative bacterial infection. Journal of Clinical Investigation. 2009;**119**:1921-1930. DOI: 10.1172/JCI36411

[99] Seki E, Brenner DA. Toll-like receptors and adaptors molecules in liver disease: Update. Hepatology. 2008;**48**:322-335. DOI: 10.1002/hep.22306

[100] Nathan C. Neutrophils and immunity: Challenges and opportunities. Nature Reviews Immunology. 2006;**6**:173-182. DOI: 10.1038/nri1785

[101] Zindel J, Kubes P. DAMPs, PAMPs, and LAMPs in immunity and sterile inflammation. Annual Review of Pathology. 2020;**15**:493-518. DOI: 10.1146/annurev-pathmechdis-012419-032847

[102] Weiss SJ. Tissue destruction by neutrophils. New England Journal of Medicine. 1989;**320**:365-376. DOI: 10.1056/NEJM198902093200606

[103] Henson PM, Johnston RB Jr. Tissue injury in inflammation. Oxidants, proteinases, and cationic proteins. Journal of Clinical Investigation. 1987;**79**:669-674. DOI: 10.1172/JCI112869

[104] Duffield JS, Ware CF, Ryffel B, Savill J. Suppression by apoptotic cells defines tumour necrosis factor-mediated induction of glomerular mesangial cell apoptosis by activated macrophages. American Journal of Pathology. 2001;**159**:1397-1404. DOI: 10.1016/S0002-9440(10)62526-6

[105] Basil MC, Levy BD. Specialized pro-resolving mediators: Endogenous regulators of infection and inflammation. Nature Reviews Immunology. 2016;**16**:51-67. DOI: 10.1038/nri.2015.4

[106] Sugimoto MA, Sousa LP, Pinho V, Perretti M, Teixeira MM. Resolution of inflammation: What controls its onset? Frontiers in Immunology. 2016;**7**:160. DOI: 10.3389/fimmu.2016.00160

[107] Serhan CN, Savill J. Resolution of inflammation: The beginning programs the end. Nature Immunology. 2005;**6**:1191-1197. DOI: 10.1038/ni1276

[108] Clapes T, Lefkopoulos S, Trompouki E. Stress and non-stress roles of inflammatory signals during HSC emergence and maintenance. Frontiers in Immunology. 2016;**7**:487. DOI: 10.3389/fimmu.2016.00487

[109] Takizawa H, Boettcher S, Manz MG. Demand-adapted regulation of early hematopoiesis in infection and inflammation. Blood. 2012;**119**:2991-3002. DOI: 10.1182/blood-2011-12-380113

[110] Essers MA, Offner S, Blanco-Bose WE, Waibler Z, Kalinke U, Duchosal MA, et al. IFNalpha activates dormant haematopoietic stem cells

in vivo. Nature. 2009;**458**:904-908. DOI: 10.1038/nature07815

[111] Sato T, Onai N, Yoshihara H, Arai F, Suda T, Ohteki T. Interferon regulatory factor-2 protects quiescent hematopoietic stem cells from type I interferon-dependent exhaustion. Nature Medicine. 2009;**15**:696-700. DOI: 10.1038/nm.1973

[112] Baldridge MT, King KY, Boles NC, Weksberg DC, Goodell MA. Quiescent haematopoietic stem cells are activated by IFN-gamma in response to chronic infection. Nature. 2010;**465**:793-797. DOI: 10.1038/nature09135

[113] Pronk CJ, Veiby OP, Bryder D, Jacobsen SE. Tumor necrosis factor restricts hematopoietic stem cell activity in mice: Involvement of two distinct receptors. Journal of Experimental Medicine. 2011;**208**:1563-1570. DOI: 10.1084/jem.20110752

[114] Yamazaki S, Ema H, Karlsson G, Yamaguchi T, Miyoshi H, Shioda S, et al. Nonmyelinating Schwann cells maintain hematopoietic stem cell hibernation in the bone marrow niche. Cell. 2011;**147**:1146-1158. DOI: 10.1016/j.cell.2011.09.053

[115] Zhao JL, Ma C, O'Connell RM, Mehta A, DiLoreto R, Heath JR, et al. Conversion of danger signals into cytokine signals by hematopoietic stem and progenitor cells for regulation of stress-induced hematopoiesis. Cell Stem Cell. 2014;**14**:445-459. DOI: 10.1016/j.stem.2014.01.007

[116] Nagai Y, Garrett K, Ohta S, Bahrun U, Kouro T, Akira S, et al. Toll-lile receptors on hematopoietic stem cells stimulate innate immune system replenishment. Immunity. 2006;**24**:801-812. DOI: 10.1016/j.immuni.2006.04.008

[117] Kim PG, Canver MC, Rhee C, Ross SJ, Harriss JV, Tu HC, et al. Interferon-α signaling promotes embryonic HSC maturation. Blood. 2016;**128**:204-216. DOI: 10.1182/blood-2016-01-689281

[118] Giorgetti A, Castano J, Bueno C, Diaz de la Guardia R, Delgado M, Bigas A, et al. Proinflammatory signals are insufficient to drive definitive hematopoietic specification of human HSCs in vitro. Experimental Hematology. 2017;**45**:85-93.e2. DOI: 10.1016/j.exphem.2016.09.007

[119] Pietras EM, Mirantes-Barbeito C, Fong S, Loeffler D, Kovtonyuk LV, Zhang S, et al. Chronic interleukin-1 exposure drives haematopoietic stem cells towards precocious myeloid differentiation at the expense of self-renewal. Nature Cell Biology. 2016;**18**:607-618. DOI: 10.1038/ncb3346

[120] Kohchi C, Noguchi K, Tanabe Y, Mizuno D, Soma G. Constitutive expression of TNF-alpha and -beta genes in mouse embryo: Roles of cytokines as regulator and effector on development. International Journal of Biochemistry. 1994;**26**:111-119. DOI: 10.1016/0020-711x(94)90203-8

[121] Espin-Palazon R, Stachura DL, Campbell CA, Garcia-Moreno D, Del Cid N, Kim AD, et al. Proinflammatory signaling regulates hematopoietic stem cell emergence. Cell. 2014;**159**:1070-1085. DOI: 10.1016/j.cell.2014.10.031

[122] Helbing T, Arnold L, Wiltgen G, Hirschbihl E, Gabelmann V, Hornstain A, et al. Endothelial BMP4 regulates leukocyte diapedesis and promotes inflammation. Inflammation. 2017;**40**:1862-1874. DOI: 10.1007/s10753-017-0627-0

[123] Jo H, Song H, Mowbray A. Role of NADPH oxidases in disturbed flow-and

BMP4-induced inflammation and atherosclerosis. Antioxidants & Redox Signaling. 2006;**8**(9-10):1609-1619. DOI: 10.1089/ars.20 06.8.1609

[124] Zhao X, Zhang J, Zhang W, Dai R, Xu J, Li Z, et al. The relationship between circulating bone morphogenetic protein-4 and inflammation cytokines in patients undergoing thoracic surgery: A prospective randomized study. Journal of Inflammation Research. 2021;**14**:4069-4077. DOI: 10.2147/JIR.S324775

Chapter 5

Immune Cell Generation from Human-Induced Pluripotent Stem Cells: Current Status and Challenges

Yu-Yun Xiong and Yun-Wen Zheng

Abstract

The immune system plays a crucial role in recognizing and eliminating foreign antigens, working in conjunction with other bodily systems to maintain the stability and physiological balance of the internal environment. Cell-based immunotherapy has revolutionized the treatment of various diseases, including cancers and infections. However, utilizing autologous immune cells for such therapies is costly, time-consuming, and heavily reliant on the availability and quality of immune cells, which are limited in patients. Induced pluripotent stem cell (iPSC)-derived immune cells, such as T cells, natural killer (NK) cells, macrophages, and dendritic cells (DCs), offer promising opportunities in disease modeling, cancer therapy, and regenerative medicine. This chapter provides an overview of different culture methods for generating iPSC-derived T cells, NK cells, macrophages, and DCs, highlighting their applications in cell therapies. Furthermore, we discuss the existing challenges and future prospects in this field, envisioning the potential applications of iPSC-based immune therapy.

Keywords: NK cells, macrophage, iPSC-derived cells, cellular therapy, T cells

1. Introduction

Immunotherapy has emerged as a highly promising therapeutic approach, particularly in the field of anticancer treatment, showcasing remarkable clinical efficacy. It encompasses various strategies, such as adoptive cell transfer (ACT) and immune checkpoint inhibitors (ICIs). Immune cells, being central players in the pathogenesis and progression of numerous diseases, serve as the fundamental components underlying the effectiveness of immunotherapy [1, 2]. Currently, a Phase I/IIa clinical trial (NCT03666000) is underway, investigating the efficacy of allogeneic CD19 chimeric antigen receptor (CAR)-T cell therapy in the treatment of relapsed/refractory B-non-Hodgkin lymphoma (NHL) and B-cell acute lymphoblastic leukemia (ALL). Furthermore, a Phase I clinical trial (NCT04220684) is evaluating the potential of allogeneic natural killer (NK) cell therapy for acute myeloid leukemia (AML). However, the use of autologous immune cells in such immunotherapy approaches is associated with significant drawbacks, including high costs, time-intensive procedures, and a heavy reliance on the availability and quality of immune cells, which are limited in patients. Consequently, there is an urgent need to develop an "off the shelf" strategy in an allogeneic setting, allowing for the unlimited proliferation of immune cells to address these challenges and facilitate clinical advancements.

In 2006, the Takahashi and Yamanaka research group achieved a groundbreaking milestone by generating induced pluripotent stem cells (iPSCs) from fibroblasts through the transfection of key factors known as the "Yamanaka factors." This pioneering approach involved the introduction of OCT3/4, SOX2, KLF4, MYC, NANOG, and LIN28 into the cells, resulting in their reprogramming into a pluripotent state [3, 4]. These iPSCs possess similar properties to those of embryonic stem cells (ESCs) in terms of morphology, growth characteristics, and developmental potential [5, 6]. iPSCs offer distinct advantages in the production of immunotherapeutic cells, primarily due to their ability to undergo unlimited reproduction *in vitro* and their ease of genetic modification. These characteristics make iPSCs highly valuable in the generation of immunotherapeutic cells for various applications [7, 8]. Furthermore, unlike ESCs, the clinical use of iPSCs does not raise ethical concerns. As a result, iPSC technology holds great potential for the development of allogeneic "off-the-shelf" cellular therapeutics that can benefit a larger number of patients. This approach is anticipated to be the most effective method for treating various types of malignancies. The advent of iPSC-derived immune cells signifies the beginning of a new era in immunotherapy, paving the way for innovative advancements in the field [9]. In this chapter, we will provide an overview of the key methods utilized for generating immune cells from iPSCs, along with their potential clinical applications and inherent limitations.

2. iPSCs-T cells

T lymphocytes are derived from hematopoietic stem cells (HSCs) located within the bone marrow. Following their production in the bone marrow, hematopoietic progenitors migrate to the thymus, where they undergo maturation into functional T cells under the influence of thymic hormones. Matured T cells are then distributed throughout the body via the bloodstream, reaching thymus-dependent regions of peripheral immune organs. They can also circulate through lymphatic vessels, peripheral blood, and tissue fluid, performing essential functions in cellular immunity and immune regulation [10]. Adoptive T-cell immunotherapy has emerged as a promising therapeutic strategy for treating various cancers and viral infections [11–14]. However, the current processes involved in generating T-cell lines from donors or genetically modifying autologous T cells for each patient are time-consuming and expensive. These limitations hinder the widespread and convenient utilization of T cells with antigen specificity. Moreover, the exhaustion of antigen-specific T cells remains a significant challenge in this approach. There is an urgent need for an unlimited supply of T lymphocytes with antigen-specific characteristics to enhance the effectiveness of T-cell therapies. In this regard, potential sources for such T cells include peripheral blood T cells from healthy donors and T cells generated from iPSCs. iPSC technology, as an "off-the-shelf" source of T cells, holds the potential to generate antigen-specific T cells that not only fulfill the requirements for large-scale clinical applications but also ensure the expression of identical T-cell receptor (TCR) genes [15, 16].

2.1 Generation

In 2002, Hochedlinger and Jaenisch conducted a groundbreaking experiment, demonstrating the successful transfer of mature lymphocyte nuclei into oocytes. This pioneering approach enabled them to establish ESCs from cloned blastocysts. To advance their research, they further injected these ESCs into tetraploid blastocysts, leading to the generation of monoclonal mice. These significant findings provided compelling evidence that a fully differentiated cell possesses the capacity for reprogramming and can give rise to an adult cloned animal [17]. In nearly all protocols for iPSC-T cell differentiation, the initial step involves reprogramming T-iPSCs from sorted $CD4^{+}$ helper T cells and $CD8^{+}$ cytolytic T cells obtained from a healthy donor. This reprogramming process is typically achieved through the transfection of Sendai virus vectors or episomal plasmid vectors carrying the four Yamanaka factors [15, 18–21]. During the process of reprogramming, it is possible to retain the antigenic specificity of T cells. This is because T-iPSCs inherit the same rearranged TCR genes at the T-cell receptor loci as the original T cells, which allows for the generation of functional T cells with the desired antigenic specificity [15, 22].

To differentiate T-iPSCs into iPSC-T cells, three different methods have been employed. These methods include the two-dimensional (2D) Delta-like ligand (DLL)1/DLL4-expressing stroma system [22–26], 2D stroma-free system [27, 28], and three-dimensional (3D) artificial thymus organoid (pluripotent stem cell-artificial thymus organoid (PSC-ATO)) system [29, 30]. In 2013, three separate research groups pursued a similar approach by employing feeder cells to differentiate iPSC-T cells. Typically, T-iPSCs were initially cultured on mouse embryonic feeder (MEF) cells to expand pluripotent stem cells *ex vivo*. Subsequently, they were redifferentiated into hematopoietic progenitors using OP9 or C3H10T1/2 feeder cells. It is worth noting that the Notch pathway plays a critical role in the generation of HSCs during this process [31–33]. In mammals, the Notch signaling pathway comprises four Notch receptors (Notch 1–4) and five Notch ligands, including Delta-like ligands (Dll) 1, 3, and 4, and Jagged 1 and 2 (JAG 1 and 2). This pathway is highly conserved and plays a crucial role in various developmental processes, particularly in hematopoiesis [34, 35]. Dar Heinze et al. conducted a study where they discovered that early stimulation of the Notch pathway, achieved through the use of OP9-hDLL4 feeder cells or hDLL4-coated plates, directed hematopoietic progenitors toward differentiation into NK cells and T cells. This finding highlights the critical role of Notch signaling in guiding the fate determination of hematopoietic progenitors toward these specific immune cell lineages [36]. The final step in the differentiation of iPSC-T cells involved seeding the cells onto OP9-DL1 feeder cells and utilizing a combination of cytokines to promote the production of functional T cells [23, 24, 26]. This standardized approach proved successful in generating a significant number of $CD8^{+}$ T cells, with over 90% of these cells originating from the same T-iPSC source. The results demonstrated that iPSCs are a potent tool for generating and developing T-cell lineages *in vitro*. This advancement holds great potential in the field of regenerative medicine, particularly for the progress of allogeneic therapies [37].

Feeder cells were employed in the generation of iPSC-derived T cells at various stages of the process. Nevertheless, the use of murine-derived stroma feeder layers raises concerns about potential cross-species contamination [27, 38]. Moreover, the utilization of different feeder cells necessitates distinct combinations of serum and basal media for maintenance culture. This complexity in culturing conditions can increase the risks of uncontrolled differentiation and pose challenges for ensuring quality control of feeder cells and serum. To address this issue, Iriguchi et al. devised a feeder-free and serum-free culture system for differentiating iPSC-T cells [27]. First, they induced hematopoietic progenitors from T-iPSC cell lines. Embryoid bodies (EBs) were generated from single cells without the need for feeder cells and serum. Next, hematopoietic cells were induced in the presence of CHIR99021, bone morphogenetic protein 4 (BMP-4), basic fibroblast growth factor (bFGF), and vascular endothelial growth factor (VEGF), and they underwent proliferation in the presence of hematopoietic cytokines, specifically SB431542. Second, to enhance the production of T cells, a combination of CXCL12-SDF1α and a p38 inhibitor, SB203580, was utilized to generate $CD4^{+}CD8\alpha\beta^{+}$ double-positive (DP) cells. This innovative approach demonstrates remarkable efficiency and scalability in generating fully functional CD8αβ T cells from iPSCs. More recently, a 3D organoid culture system was reported to successfully generate CAR T cells for "off-the-shelf" manufacturing strategies [29]. Montel-Hagen et al. introduced a 3D artificial thymus organoid (ATO) culture system for the *in vitro* differentiation of human hematopoietic stem and progenitor cells (HSPCs) into functional, mature T cells. They achieved this by utilizing a standardized stromal cell line expressing Notch ligands in a serum-free environment [39]. This innovative continuous culture system facilitated both the specification of hematopoietic cells and their subsequent terminal differentiation into naïve $CD3^{+}CD8\alpha\beta$ + and $CD3^{+}CD4^{+}$ conventional T cells.

While the redifferentiation strategy allows for the realization of "off-the-shelf" T cells [8, 40, 41], there is a risk of rejection by the patient's immune system when using allogeneic T cells. Additionally, a significant concern arises from the production of heterogeneous T cells, which can lead to the generation of potentially harmful alloreactive T cells at varying frequencies. In order to circumvent immune rejection and the production of polyclonal T cells, T-iPSCs can also be derived from patients themselves. Daopeng Yang conducted a study where T-iPSCs were generated from cytotoxic T lymphocytes infiltrating hepatocellular carcinoma (HCC). This was achieved using an integrative Sendai virus vector. The resulting pluripotent cell line exhibited a normal karyotype and could be redifferentiated into rejuvenated CTLs specifically targeting HCC [42]. Munenari Itoh conducted a study where T-iPSCs were generated from monocytes of a melanoma patient. $CD8^{+}$ T cells were sorted after stimulation with tumor antigens, and then reprogrammed into iPSCs through the exogenous expression of reprogramming factors, utilizing the Sendai virus vector [43].

The generation of T lymphocytes from induced pluripotent stem cells (iPSCs) *in vitro* holds promise for adoptive T-cell therapy. However, the yield and efficiency of lymphoid cells have been limited, and their properties are still only partially understood [7, 44]. Studying T cells derived from ESCs and iPSCs faces challenges due to a limited understanding of their antigen specificity and human leukocyte antigen (HLA) restriction. T cells generated in the laboratory from ESCs or iPSCs exhibit unpredictable T-cell receptor (TCR) repertoires due to random rearrangements of TCR genes, and the mechanisms involved in their selection during *in vitro* differentiation are not yet well understood. Nevertheless, this limitation can be overcome by utilizing iPSCs that possess an endogenous TCR with known antigen specificity [23, 24]. However, this approach requires a time-consuming procedure of cloning T cells specific to the antigen, and it is limited to antigens that can be identified using unique T cells from each patient. Additionally, the clinical use of T cells that recognize antigens through their inherent TCR is constrained by the need to match their specificity with the HLA molecules of the recipient patient.

2.2 Translation

One of the notable advantages of iPSCs is their versatility in genetic modification. iPSC-derived T cells have emerged as a prominent tool in cancer immunotherapy. By introducing a chimeric antigen receptor (CAR) or transgenic T-cell receptor (TCR) gene, iPSCs can be transformed into antigen-specific T cells capable of effectively targeting and eliminating cancer cells *in vitro* and *in vivo* [23, 24, 26]. The incorporation of CRISPR-Cas9 (clustered regularly interspaced short palindromic repeats-CRISPR-associated protein 9) technology further enhances the efficacy and safety of this approach. Through CRISPR-Cas9-mediated insertion of CAR genes into the TRAC locus of the endogenous TCRα constant (TRAC) gene, the risk of host-related alloreactions can be minimized, thus improving immune compatibility [45]. Genetic editing of iPSCs has proven to be a valuable strategy in broadening immune compatibility, supporting the potential for developing "off-the-shelf" cell products. In the context of T-iPSCs (T cell-derived iPSCs), reprogrammed iPSC clones can inherit the original TCR and subsequently be redifferentiated into iPSC-T cells (T cells redifferentiated from iPSCs). However, this approach necessitates time-consuming cloning of antigen-specific T cells and is limited to antigens that can be identified from patient-specific T cells. Furthermore, the therapeutic application of iPSC-T cells is restricted by the need for HLA compatibility between the donors and recipients, significantly limiting the universality of potential "off-the-shelf" applications [16, 25].

Engineering strategies for iPSC-derived T cells involve the introduction of TCR and CAR constructs. TCR-mediated therapies using iPSC-T cells have been carried out by either inheriting the endogenous TCR genes from antigen-specific T cells [22–24] or genetically incorporating exogenous TCRs [25, 29]. Clinical trials have shown the effectiveness of TCR-engineered primary T cells targeting melanoma antigen MART1 [46] and germline antigen New York esophageal squamous cell carcinoma 1 (NY-ESO1) [47]. In these studies, TCR-engineered iPSC-T cells have demonstrated promising results in clinical settings, validating their potential as therapeutic agents for cancer treatment.

Successful outcomes have been observed when introducing a CAR into iPSCs or directly into iPSC-derived T cells to generate antigen-specific T cells [26–28, 30, 48]. CAR engineering effectively redirects T-cell specificity in an HLA-independent manner, eliminating the need for HLA restriction and enhancing the antitumor properties. In a study conducted by Themeli et al. in 2013, T-iPSC clones were generated by reprogramming peripheral T cells from a healthy donor. Subsequently, a second-generation CAR specific for CD19 was transduced into the selected T-iPSC clone. The resulting CAR-expressing iPSC-T cells exhibited remarkable antitumor efficacy in a xenograft model, although they exhibited phenotypic similarities to innate γδT cells [26]. This method effectively generated an innate type of T cell expressing a CD8αα homodimer, which influenced the antigen-specific cytotoxic capacity of the redifferentiated T cells in a manner akin to MART-1-specific T cells. To improve upon the conventional approach, Maeda et al. modified the method by purifying differential pressure (DP) cells, which were subsequently stimulated with monoclonal anti-CD3 antibodies to generate CD8αβ T cells. These modified T cells displayed comparable antigen-specific cytotoxicity to the original cytotoxic T lymphocytes (CTLs) [15].

In recent years, iPSC-T cells have emerged as a promising approach in the treatment of various diseases. One notable example is their application in highly aggressive lymphoma, specifically extranodal NK/T-cell lymphoma of the nasal type (ENKL). iPSC-derived cytotoxic T lymphocytes, specifically designed to target the EBV antigen, have exhibited remarkable results. These cytotoxic T lymphocytes have significantly prolonged patient survival, demonstrated potent tumor-suppressive effects, and persisted as central memory T cells *in vivo* for a minimum of 6 months [49–55]. Rejuvenated cytotoxic T lymphocytes have also exhibited promising outcomes in the treatment of various solid tumors, including cervical cancer associated with human papillomavirus (HPV) infection [56] and renal cell carcinoma [57]. For instance, in cervical cancer cases, T-iPSCs derived from human papillomavirus type 16 (HPV16) E6- or E7-specific cytotoxic T lymphocytes efficiently differentiated into rejuvenated CTLs that specifically targeted HPV16. These rejuvenated CTLs not only displayed prolonged survival compared to the original CTLs but also induced increased tumor shrinkage and significantly prolonged survival in an *in vivo* mouse model [56]. Similarly, in renal cell carcinomas, T-iPSCs established from Wilms tumor 1-specific CTLs effectively suppressed tumor growth in an *in vivo* mouse model [57]. These findings underscore the potential of iPSC-T cells as a valuable therapeutic strategy for targeting and treating various types of cancers, opening new avenues for therapeutic interventions.

3. iPSCs-NK cells

Natural killer cells are a type of lymphocyte that originates from common lymphoid progenitors (CLPs) during hematopoiesis [58]. They are predominantly found in the bone marrow, peripheral blood, liver, spleen, lung, and lymph nodes. Unlike T and B cells, NK cells possess the unique ability to eliminate tumor cells and virus-infected cells without prior sensitization. NK cells play a crucial role in immune responses, including antitumor activity, defense against viral infections, immune regulation, and even involvement in hypersensitivity and autoimmune diseases in certain cases. They possess the ability to recognize target cells and mediate their killing. When compared to T cells, allogeneic NK cell transfers have a lower risk of graft-versus-host disease (GVHD) and may even decrease the overall risk [59, 60]. However, current strategies dependent on donor cells can only provide a limited supply of custom-made therapeutic NK cells for a restricted number of patients. To overcome this limitation and offer a more accessible treatment option, induced pluripotent stem cells (iPSCs) have been employed in immunotherapy to enable the mass production of NK cells from iPSCs. This approach holds the potential to provide an unlimited supply of "off-the-shelf" NK cells, benefiting a larger number of recipients.

3.1 Generation

In contrast to the redifferentiation process of iPSCs into iPSC-T cells, NK cells can be reprogrammed directly from iPSC cell lines [61–63] or generate from peripheral blood cells [64] and human fibroblasts [65]. The iPSC cell lines used in these approaches were established from various sources, including umbilical cord blood CD34+ cells and newborn human foreskin fibroblasts [66, 67].

Two methods for the production of iPSC-derived NK cells can be categorized based on the use of feeder cells. In the standard protocol, iPSC cell lines were cultured on MEFs and differentiated into hematopoietic progenitors using M210-B4 cells. To generate spin EBs suitable for aggregation, iPSCs were passaged in TrypLE Select on a low-density MEF layer. Subsequently, the spin EBs were seeded onto plates either with or without EL08-1D2 (a murine embryonic liver cell line) for NK cell differentiation.

This differentiation process was carried out in the presence of specific NK cell initiating cytokines, including interleukin (IL)-3, IL-7, IL-15, stem cell factor (SCF), and fms-like tyrosine kinase receptor-3 ligand (FLT3L) [62, 63, 68].

A method developed by Frank Cichocki et al. eliminates the need for spin EB generation. In this approach, iPSCs were cultured in a combination of small molecules and cytokines to generate CD34$^+$ hematopoietic progenitor cells. These CD34$^+$ cells were then cocultured with stromal cells that were transduced with Notch ligand and supplemented with cytokines that support the proliferation and differentiation of hematopoietic progenitor cells toward the NK cell lineage. Subsequently, the cells were cocultured with modified K562 cells to further expand the differentiation of iPSC-derived NK cells. This method offers a streamlined process for the efficient production and expansion of iPSC-NK cells [65].

In 2021, Kyle B Lupo et al. developed a serum- and feeder-free system for differentiating iPSCs into NK cells [69]. The differentiation process involved several key steps. First, iPSCs were differentiated into hematopoietic cells using a hematopoietic differentiation medium comprising STEMdiff APEL 2, SCF, bone morphogenetic protein 4, vascular endothelial growth factor, and a rho-associated protein kinase (Rock) inhibitor. To facilitate the formation of EBs, the cells were subjected to a spinning step. After 11 days, hematopoietic progenitor cells were collected from the EBs and seeded in a specialized NK cell differentiation medium containing STEMdiff APEL 2, SCF, IL-7, IL-15, and FLT3L to initiate the differentiation into NK cells. This novel system offers a serum- and feeder-free approach for efficient and controlled differentiation of iPSCs into functional NK cells.

3.2 Translation

Natural killer cells derived from iPSCs offer the advantage of not being HLA restricted, making iPSC-NK cells an excellent candidate for allogeneic "off-the-shelf" immunotherapy [66]. These iPSC-NK cells serve as a readily available source of cells for immunotherapy, capable of targeting tumors and activating the adaptive immune system to transform a "cold" tumor into a "hot" one by facilitating the recruitment of activated T cells, thus enhancing the efficacy of checkpoint inhibitor therapies [65]. The ability to produce iPSC-NK cells under defined conditions and their demonstrated functional responses indicate their potential as effective therapeutic agents in adoptive transfer settings for treating solid tumors. They offer a renewable source of donor-independent NK cells for immunotherapy, holding great promise in clinical applications [69].

Moreover, iPSC-NK cells, being derived from iPSCs, possess the characteristic feature of being amenable to genetic editing. One strategy for genetic modification of iPSC-NK cells is the incorporation of CARs to enhance their antitumor cytotoxicity. Reports have shown that CARs effectively reprogram NK cell specificity [70]. Notably, Laurent Boissel et al. observed that CAR-NK cells exhibited enhanced elimination of primary chronic lymphocytic leukemia (CLL) cells through antibody-dependent cell-mediated cytotoxicity (ADCC) mediated by anti-CD20 monoclonal antibodies [71]. iPSC-NK cells engineered with CARs offer several advantages: (1) they have fewer complications such as cytokine release syndrome (CRS), neurotoxicity, or GVHD; (2) they are not restricted by HLA; and (3) they can activate cytotoxic effects independently of the CAR itself [72–77]. In a study by Dan Kaufman's group, a first-generation CAR incorporating CD4/CD3ζ was introduced into iPSC-NK cells, demonstrating their ability to suppress human immunodeficiency virus (HIV) replication in $CD4^+$ T cells [68]. Furthermore, Li et al. tested a series of specialized CARs incorporating costimulatory molecule intracellular domains and found that iPSC-derived NK cells expressing CAR (NK-CAR-iPSC-NK cells) exhibited a typical NK cell phenotype and demonstrated superior antitumor activity compared to iPSC-derived NK cells expressing T-cell CARs (T-CAR-iPSC-NK cells) or non-CAR-expressing cells, both *in vitro* and *in vivo* [78].

The persistence and enhanced functional activity of NK cells rely on their interaction with various immune cells that release different cytokines. Among these cytokines, IL-15 plays a crucial role in the differentiation of NK cells [79]. However, during *in vitro* culturing, the frequent addition of IL-15 is necessary due to its short half-life [80]. To overcome this limitation, innovative approaches have been explored, including the use of IL-15 constructs such as secreted IL-15 or an IL-15/IL-15-receptor fusion construct (IL-15RF). Woan et al. developed triple-gene-edited iPSC-NK cells with a high-affinity, noncleavable version of the Fc receptor CD16a, a membrane-bound interleukin (IL)-15/IL-15R fusion protein, and a knockout of the ecto-enzyme CD38, which hydrolyzes NAD^+. They discovered that these engineered iPSC-NK cells exhibited enhanced anticancer effects in leukemia and multiple myeloma [81]. Another important negative regulator of IL-15 signaling in NK cells is cytokine-inducible SH2-containing protein (CIS), encoded by the CISH gene. Huang Zhu et al. found that knockout of CISH in iPSC-NK cells improved the expansion capacity of NK cells and increased their cytotoxic activity against multiple tumor cell lines when maintained at low cytokine concentrations [82]. These modified IL-15 forms provide sustained proliferation signals, thereby augmenting the antitumor efficacy of NK cells both in laboratory settings and in living organisms [83, 84].

Antibody-dependent cell-mediated cytotoxicity is a mechanism by which NK cells exert cytotoxicity through the Fc receptor CD16a. However, CD16a has a low affinity for tumor-bound IgG antibodies and is susceptible to cleavage by a disintegrin and metalloprotease 17 (ADAM17) upon NK cell activation. To address these limitations, Kristin M Snyder et al. enhanced the binding ability of NK cells to antitumor monoclonal antibodies (mAbs) by constructing a fusion protein comprising CD64, the highest-affinity Fc-gamma receptor (FcγR) expressed by leukocytes, and CD16A. This CD64/16A fusion protein lacked the ADAM17 cleavage region in CD16A, preventing downregulation of expression following NK cell activation during ADCC. The CD64/16A iPSC-NK cells exhibited enhanced conjugation to antibody-treated tumor cells, improved ADCC, cytokine production, and ultimately mediated effective tumor cell killing [85]. Another strategy involves mutating CD16a to produce a high-affinity noncleavable variant known as hnCD16. When hnCD16 was incorporated into iPSC-NK cells, the resulting hnCD16-iPSC-NK cells exhibited functional maturity and demonstrated enhanced ADCC against multiple tumor targets. In *in vivo* xenograft studies using a human B-cell lymphoma model, the combination of hnCD16-iPSC-NK cells and anti-CD20 monoclonal antibodies significantly improved regression of B-cell lymphoma and increased overall survival [86]. Additionally, Fanyi Meng fused the ectodomain of hnCD16 with NK cell-specific activating domains in the cytoplasm. This fusion protein showed improved ADCC and cytotoxicity *in vitro* and *in vivo*, as observed in coculture experiments with tumor cell lines and in a xenograft mouse model bearing human B-cell lymphoma [87].

As mentioned above, iPSC-NK cell technology has been utilized for the treatment of hematologic malignancies. However, its applications extend beyond that and encompass the field of solid tumors and viral infections as well. Studies have demonstrated the efficacy of iPSC-derived NK cells in various contexts. For instance,

Hermanson DL found that iPSC-derived NK cells enhanced the antitumor effect and prolonged survival in ovarian cancer [66]. Furthermore, iPSC-derived NK cells have shown promise as an improved approach for treating HIV infection [61, 68] and COVID-19 [88]. These findings highlight the broad potential of iPSC-NK cells in combating a range of diseases, including both cancers and viral infections.

4. iPSCs-macrophages

Macrophages are a type of white blood cells that reside within tissues and are derived from monocytes, which themselves originate from precursor cells in the bone marrow. Macrophages, along with monocytes, function as phagocytes involved in both nonspecific defense (innate immunity) and specific defense (cellular immunity) in vertebrates. Their primary role is to engulf and digest cell fragments and pathogens, whether in the form of stationary or free cells, and to activate lymphocytes or other immune cells to mount a response against pathogens. Macrophages are immune cells with diverse functions, making them crucial subjects for the study of cellular immunity and molecular immunology. These nonreproductive cells can survive for 2–3 weeks under favorable conditions. While primary cultures of macrophages are often used, they are challenging to maintain for extended periods. Immortalized macrophage cell lines are not suitable for clinical applications, and engineering bone marrow or peripheral blood mononuclear cell (PBMC)-derived primary macrophages is not efficient. Therefore, iPSC-derived macrophages represent a valuable source for myeloid cell-based immunotherapy, offering great potential in the field of immunotherapy [89].

4.1 Generation

Similar to the production of iPSC-NK cells, iPSCs utilized for differentiating into macrophages originate from iPSC cell lines derived through the reprogramming of fibroblasts using iPSC reprogramming vectors such as OCT4, SOX2, KLF4, and c-MYC [90–92], $CD34^+$ bone marrow cells [93] or peripheral blood monocytes [94, 95]. The methods employed to generate iPSC-macrophages can also be categorized based on the use of feeder cells.

In standard protocols, iPSC cells are initially cocultured with feeder cells such as OP9 mouse stromal cells, in the presence of bone morphogenetic protein 4. This culture condition leads to the differentiation of iPSCs into either 37.8% CD133 HSCs or 9–17% $CD43^+$ hematopoietic progenitors. To generate macrophages, myelomonocytic colonies are cultured with granulocyte-macrophage colony-stimulating factor (GM-CSF) and macrophage colony-stimulating factor (M-CSF). The resulting iPSC-derived macrophages exhibit functionality and, upon stimulation, secrete substantial amounts of IL-6, IL-10, and tumor necrosis factor alpha (TNF-α) compared to nonstimulated macrophages [90]. IL-3 plays a crucial role in promoting the proliferation of various types of hematopoietic cells during early primitive hematopoiesis and definitive hematopoietic specification. Lachmann, N. et al. combined the use of IL-3 with M-CSF or G-CSF to achieve prolonged and large-scale production of functional granulocytes as well as monocytes/macrophages through EB-based hematopoietic *in vitro* differentiation [96]. They initiated EB formation in ESC medium supplemented with basic fibroblast growth factor (bFGF) and a Rock inhibitor. Subsequently, an intermediate myeloid-cell-forming complex (MCFC) was generated by culturing the EBs in albumin polyvinylalcohol essential lipid (APEL) medium supplemented with human IL-3, human M-CSF, human G-CSF, or human GM-CSF for a period of 7 days. From day 10 to day 15 onward, monocytes/macrophages or granulocytes were generated. To further promote maturation, the generated monocytes/macrophages or granulocytes were cultured in Roswell Park Memorial Institute (RPMI) 1640 Medium supplemented with 10% fetal serum, L-glutamine, human M-CSF, human G-CSF, or human GM-CSF for 7–10 days.

The use of feeder cells or serum in the culture system adds additional biological and regulatory complexities, which may limit the clinical utility of iPSC-derived monocytes and macrophages. To overcome these challenges, a fully chemically defined, serum- and feeder-free protocol has been developed, significantly improving reproducibility [97, 98]. In this protocol, iPSC cells are used to generate spin embryoid bodies (EBs) in a culture medium supplemented with BMP4, VEGF, and SCF. Subsequently, the EBs are collected and passed through a 40-μm strainer before being transferred to a "factory" medium. This medium, known as X-VIVO 15, is supplemented with Glutamax, 1% penicillin/streptomycin, mercaptoethanol, M-CSF, and IL-3. Utilizing this serum-free protocol, differentiation cultures are established, which continue to produce harvestable and uniform monocytes for extended periods, often lasting up to 1 year [98]. López-Yrigoyen, M. also successfully generated macrophages from an iPSC line using this method [99].

During the differentiation process, the generation of EBs typically involves reseeding and size control steps. However, alternative protocols modified by Cao et al. have been developed to eliminate the need for EB generation by employing serum-free culture conditions. Despite this improvement, the yield obtained from these protocols is relatively low, thereby limiting the scalability of the studies [100]. To address this limitation, Cui, D. et al. present a fully optimized differentiation protocol that incorporates precise timing of steps and the addition of specific cytokines, chemokines, or chemicals. This optimized protocol enables large-scale production of macrophages under serum- and feeder-free conditions without the need for the EB generation step [101]. The development of this fully optimized differentiation protocol represents a significant advancement in the field, providing a reliable and scalable method for generating macrophages from iPSCs. By incorporating precise timing and additional factors, this protocol enhances the efficiency and yield of macrophage production, enabling large-scale studies and expanding the potential applications of iPSC-derived macrophages.

4.2 Translation

Induced pluripotent stem cell (iPSC)-derived macrophages can be engineered with CARs through gene modification. Klichinsky et al. engineered human primary macrophages with an anti-CD19 CAR containing a CD3ζ intracellular domain. These CAR-macrophages exhibited M1-like pro-inflammatory phenotypes and were resistant to the immunosuppressive effects of the tumor microenvironment (TME) through stimulation by the adenovirus vector [102]. Zhang et al. established a platform to engineer iPSCs with a CAR and differentiate them into macrophages, referred to as CAR-iPSCs-macrophages [89]. CAR expression conferred antigen-dependent macrophage functions, including cytokine expression and secretion, polarization toward a pro-inflammatory/antitumor state, enhanced phagocytosis of tumor cells, and demonstrated *in vivo* anticancer activity [89]. Fusing a CD20 single-chain variable fragment (scFv) to FcγR1 in iPSC-macrophages enhanced their ability to engulf and eliminate B-cell leukemic cells both *in vitro* and *in vivo* [103]. Zhang et al. successfully developed iPSC-derived CAR macrophages expressing either a CD19-specific or a mesothelin-specific fusion receptor, utilizing two distinct endodomain configurations [89]. These iPSC-derived CAR macrophages exhibited antigen-dependent anticancer functions, including cytokine expression and secretion, polarization toward a pro-inflammatory/antitumor state, enhanced phagocytosis of tumor cells, and demonstrated anticancer activity *in vivo* [89].

Macrophages are terminally differentiated cells with limited capacity for expansion. To overcome this limitation and enable the large-scale clinical use of iPSC-derived macrophages, Azusa Miyashita et al. employed gene transduction techniques to introduce genes involved in cell growth or senescence suppression, such as c-MYC, in combination with BMI1, murine double minute (MDM2), or enhancer of zeste homolog 2 (EZH2). This approach resulted in the production of human iPSC-derived macrophages that could be propagated for extended periods, functioning as primary macrophages [104]. The engineered cell lines demonstrated a low risk of tumorigenicity, as they exhibited cytokine-dependent proliferation *in vitro*. Importantly, the cytokine-rich conditions required for their growth could not be replicated in the physiological environment *in vivo*.

Patient-derived iPSCs provide valuable cellular models for studying disease pathogenesis and evaluating potential treatments. iPSC-derived macrophages, in particular, hold great promise for investigating various diseases, including cancer. Both macrophages derived from patients' monocytes and iPSCs derived from patients' fibroblasts have been utilized as models in diseases such as Gaucher disease [93]. In Gaucher disease, iPSC-derived macrophages generated from fibroblast lines obtained from patients with type 1 or type 2 Gaucher disease displayed similar characteristics. These macrophages exhibited reduced glucocerebrosidase activity and increased accumulation of glucocerebrosidase and glucosylsphingosine in lysosomes, mirroring the observations in patient monocytes. Furthermore, all the macrophages demonstrated effective phagocytosis of bacteria but exhibited reduced production of intracellular reactive oxygen species (ROS) and impaired chemotaxis [91]. Another example involves iPSCs derived from a patient with hereditary pulmonary alveolar proteinosis. When differentiated into macrophages, these cells exhibited defects in GM-CSF-dependent functions, characteristic of the disease phenotype [93].

5. iPSCs-DCs

Dendritic cells (DCs) are derived from myeloid pluripotent hematopoietic stem cells and undergo differentiation through two main pathways. Myeloid dendritic cells (MDCs) are generated by stimulation with GM-CSF and differentiate from common precursor cells shared with monocytes and granulocytes. On the other hand, lymphoid dendritic cells (LDCs) or plasmacytoid dendritic cells (pDCs) arise from lymphoid stem cells and share precursor cells with T cells and NK cells. These LDCs are also known as DC2 cells. DCs exhibit widespread distribution in various tissues, including the skin, airways, and lymphatic organs, with notable heterogeneity. Consequently, different tissues have distinct names for DCs. For instance, DCs present in the basal layer of the skin epidermis and spinous cells are referred to as Langerhans cells. DCs are considered the most potent professional antigen-presenting cells (APCs) in the body. They efficiently capture, process, and present antigens to other immune cells. Mature DCs play a crucial role in the initiation, regulation, and maintenance of immune responses by effectively activating naïve T cells, which are central to the immune response.

5.1 Generation

Dendritic cells derived from iPSCs exhibit characteristics similar to those of other immune cells. Kitadani et al. utilized dermal fibroblasts transfected with Sendai virus vectors to generate iPSCs. These iPSCs were then differentiated into hematopoietic progenitors using a combination of recombinant human bone morphogenetic protein 4 (rhBMP4), recombinant human vascular endothelial growth factor (rhVEGF), growth factor (GF), and recombinant human stem cell factor (rhSCF). Following the addition of a cytokine mixture and CD14 cell sorting, monocytic cell cultures were established and further differentiated into DC cells [105]. In contrast to Kitadani et al.'s method, a feeder-dependent system was employed for the production of iPSC-derived DCs (iPS-DCs). In this approach, iPSCs were seeded onto OP9 cell layers and cultured in the presence of GM-CSF. Once the cells differentiated into hematopoietic progenitors, the floating cells were collected and transferred to Petri dishes without feeder cells. After 5–7 days, the majority of the floating cells had differentiated into iPS-DCs. To promote their maturation, the cells were transferred and cultured in RPMI-1640/10% fetal calf serum (FCS) supplemented with GM-CSF, IL-4, TNF-α, and anti-CD40 monoclonal antibody [103, 106, 107]. These two different methods demonstrate distinct approaches for generating iPSC-derived DCs. Both techniques have proven effective in producing functional DCs from iPSCs, providing valuable tools for studying the biology of DCs and their potential applications in immunotherapy and disease modeling.

5.2 Translation

Dentritic cell vaccines have been considered as a promising option for immune cell therapy against cancer. As APCs with robust T-cell stimulating activity, DCs play a pivotal role in orchestrating the immune response. Despite their potential, clinical trials utilizing DC vaccines have encountered challenges, and the outcomes have been largely disappointing [108]. One possible reason for the limited success of DC vaccines in clinical trials is the presence of immune exhaustion in cancer patients, which compromises the ability to generate a sufficient T-cell response [109].

Dentritic cells play a crucial role in immune responses, particularly in stimulating cytotoxic T lymphocytes against viral and tumor-associated antigens through the process of cross-presentation in an MHC class I-restricted manner. Under steady-state conditions, CD141 DCs, residing in interstitial tissues, are primarily involved in maintaining immune homeostasis and inducing tolerance to local antigens. However, iPSC-derived DCs have emerged as a promising avenue for antitumor immunotherapy. In a study by Junya Kitadani et al., iPSCs derived from three healthy donors were differentiated into DCs using feeder-free culturing protocol. Carcinoembryonic antigen (CEA) complementary DNA (cDNA) was then introduced into the iPSC-derived DCs through transduction. The researchers demonstrated that these genetically modified iPSC-derived DCs were capable of inducing CEA-specific cytotoxic T lymphocytes in a human model and exhibited significant antitumor effects in a CEA transgenic mouse model [105]. Building upon their previous work, the same research group, in 2023, designed iPSC-derived DCs targeting mesothelin (MSLN) and focused on enhancing the antigen-presenting ability of these cells through the ubiquitin-proteasome system. By simultaneously expressing ubiquitin and MSLN, genetically modified iPSC-derived DCs exhibited potent cytotoxicity against tumors that naturally express MSLN, thereby overcoming immune tolerance and eliciting robust antitumor immune responses [110]. These findings highlight the potential of iPSC-derived DCs in antitumor immunotherapy. The ability to genetically modify iPSC-derived DCs to express specific antigens opens up opportunities for personalized and targeted therapies. Further research and development of iPSC-derived DC-based immunotherapies hold promise for enhancing the efficacy of cancer treatments and improving patient outcomes.

6. Limitations and challenges

Although iPSC-derived cells hold great potential for immunotherapy in clinical applications, several limitations currently hinder their widespread use. One major challenge is the low efficiency of pluripotent reprogramming across various cell types. Reprogramming adult human fibroblasts, for instance, yields a conversion rate of only 0.02–0.05% [111]. Similarly, differentiation of CD34+ mobilized human peripheral blood cells results in a conversion rate of just 0.01–0.02% [112]. In order to achieve successful cell transplantation in patients, a high yield of immune cells is required. The low reprogramming efficiency not only limits the final cell yield but also poses challenges for scaling up the process for clinical applications. Another limitation is the time-consuming nature of the differentiation process. As mentioned earlier, the differentiation of iPSCs into functional immune cells often takes 1–2 months for the development of mature properties. For iPSC-derived T cells, it typically requires 3–7 weeks to expand and reach maturity [27, 29, 38, 44]. iPSC-derived NK cell production takes at least 4 weeks [113], while iPSC-derived macrophages require around 3–7 weeks [114, 115]. The lengthy duration of expansion and differentiation not only increases costs but also prolongs the overall treatment time in clinical therapy. Furthermore, the use of murine-derived feeder cells and serum in current

	Advantages	Limitations
iPSC	• Ease of genetic editing • Unlimited source • Reduces cost • Time-consuming • Avoids ethical issue	• Low efficiency • Low functional maturity • Immunogenicity • Safety
iPSC-T cells	• Keep the antigenic specificity • Allogeneic therapies	• Immune rejection • Polyclone T cells to alloreactive • HLA restriction • More complications
iPSC-NK cells	• No HLA restriction • Less complications	• Less persistence *in vivo*
iPSC-macrophages	• Benefit to solid tumor therapy • Play a phagocytic role in tumor cells • Less toxicity and limited circulation time	• Undergo frequent polarization • Lack of proliferation capacity • Do not meet clinical safety requirements for oncogenes used

Table 1.
Advantages and limitation of induced pluripotent stem cell (iPSC)-immune cells.

culture approaches introduces the risk of cross-species contamination and variations in the final cell products. Moreover, iPSCs derived from fibroblasts pose challenges in terms of product heterogeneity. Although researchers have developed serum-free and feeder-free culture systems that are more suitable for industrial applications [69, 116], achieving standardization and consistency of the final immune cell products remains a challenge for large-scale industrial production and their use in real-patient applications and clinical trials. The safety of iPSCs is also a concern before large-scale clinical application. Recent studies have reported the tumorigenic potential of undifferentiated iPSCs and the potential for malignant transformation in differentiated iPSCs [117, 118]. Additionally, investigations into TCR gene usage in T cells derived from T-iPSCs and TCR-iPSCs have revealed a small portion of rearranged TCRs, raising further safety considerations [15].

Despite the limitations discussed above, iPSC-derived immune cells offer several advantages (see **Table 1**). To address the challenges associated with iPSC-based immunotherapy, it is crucial to develop standardized protocols and identify novel targets for cell therapy. Additionally, the influence of the immune microenvironment should be carefully considered and investigated to optimize the efficacy of iPSC-derived immune cells. By overcoming these limitations and leveraging the strengths of iPSC-based approaches, we can unlock the full potential of iPSC-immune cells in the field of immunotherapy.

7. Conclusion and future perspectives

The utilization of iPSC technology in immunotherapy has revolutionized traditional immune therapies, as it offers a virtually limitless supply of genetically engineered immune cells that can be readily available for patients' therapeutic needs. This approach eliminates the dependence on scarce cell sources from individual patients, which may not be sufficient for therapeutic purposes. By establishing iPSC banks, standardized protocols for immune cell differentiation can be implemented to ensure scalability and quality control of the generated cell products before administration. Currently, there are approximately 10 iPSC banks that have been established, with a focus on stem cell research and disease-specific cell lines, catering to the needs of both academic and industrial research endeavors [119].

The application of iPSC technology in cell therapy carries certain risks, including tumorigenicity and immune suppression. To address these concerns, novel strategies have been developed to enhance iPSC differentiation and modification. Various approaches can be employed to mitigate the tumorigenic risks associated with iPSCs. For instance, undifferentiated cells can be selectively sorted out using antibodies that target surface biomarkers [120] or eliminated through the use of cytotoxic antibodies [121]. Additionally, chemical inhibitors can be utilized to eradicate any remaining undifferentiated pluripotent cells [122, 123]. While these strategies have shown promise in reducing the risk, it is important to note that long-term culture for reprogramming and redifferentiation may still give rise to unexpected events that contribute to tumorigenicity. Consequently, caution must be exercised during the first-in-human clinical studies to anticipate and address potential issues. To enhance safety in iPSC-based cell therapies, suicide systems can be implemented as a precautionary measure. These systems are designed to induce apoptosis in transduced cells, thereby potentiating therapy without increasing toxicity or evoking cross-resistance to conventional agents. One example is the HSV-TK (herpes simplex virus thymidine kinase) gene,

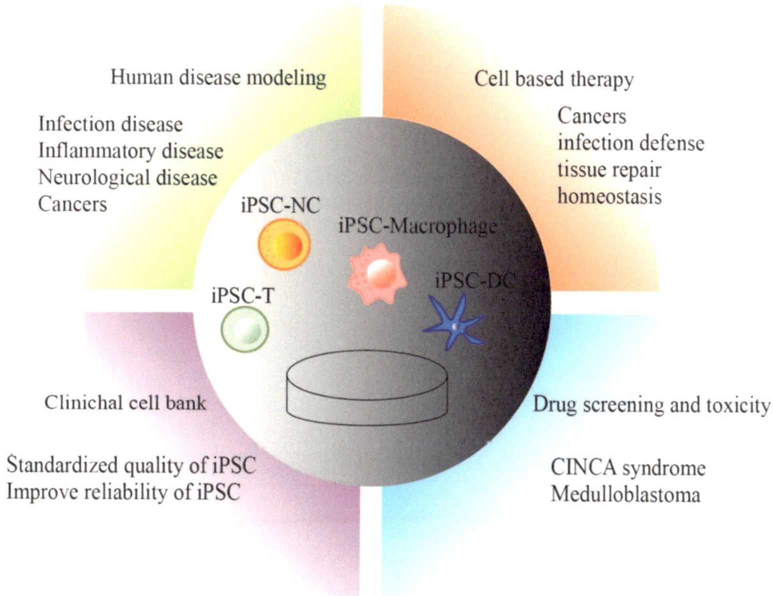

Figure 1.
Application of iPSC immune cells.

which can be combined with the administration of ganciclovir (GCV) as a safety switch in adoptive T-cell therapy or cancer treatment. However, it is important to note that certain suicide gene systems have shown limitations and may not be as clinically effective as desired [124, 125]. To safeguard against potential risks during clinical and translational investigations of iPSC-based cell therapy, Miki Ando et al. utilized the inducible caspase 9 (iC9) system as a safety mechanism. This system, consisting of an inducible caspase 9 (iC9) gene, can be activated to induce apoptosis in iPSCs if any unexpected issues arise [126].

In addition to their applications in cancer therapy, iPSC immune cells also play significant roles in establishing human disease models [91, 127, 128], drug screening, toxicity assessment [91, 97, 129], and clinical cell banking for "off-the-shelf" therapy (**Figure 1**). The versatility of iPSC immune cells enables their potential use in treating various pathological conditions beyond cancer through genetic modifications [94, 130–133]. This opens up exciting possibilities for utilizing genetically engineered iPSC immune cells as regenerative medical products in clinical practice. With further advancements and research, these cells could offer new avenues for personalized and targeted therapies in the future.

Funding statement

This research was funded partly by the National Natural Science Foundation of China (82,070,638 and 82,270,697), Jiangsu Provincial Medical Key Discipline Cultivation Unit (JSDW202229), the Science and Technology Planning Project of Guangdong Province of China (2021B1212040016), Japan Society for the Promotion of Science (JSPS), KAKENHI (JP18H02866), and the Grant for International Joint Research Project of the Institute of Medical Science, the University of Tokyo.

Author details

Yu-Yun Xiong[1] and Yun-Wen Zheng[1,2,3,4*]

1 Institute of Regenerative Medicine, Department of Dermatology, Affiliated Hospital of Jiangsu University, Jiangsu University, Zhenjiang, China

2 Guangdong Provincial Key Laboratory of Large Animal Models for Biomedicine, South China Institute of Large Animal Models for Biomedicine, School of Biotechnology and Health Sciences, Wuyi University, Jiangmen, China

3 Department of Medicinal and Life Sciences, Faculty of Pharmaceutical Sciences, Tokyo University of Science, Noda, Japan

4 Center for Stem Cell Biology and Regenerative Medicine, Institute of Medical Science, The University of Tokyo, Tokyo, Japan

*Address all correspondence to: zheng.yunwen.ld@alumni.tsukuba.ac.jp

References

[1] Zhang Y, Zhang Z. The history and advances in cancer immunotherapy: Understanding the characteristics of tumor-infiltrating immune cells and their therapeutic implications. Cellular & Molecular Immunology. 2020;**17**(8):807-821

[2] Liu Y, Yan X, Zhang F, Zhang X, Tang F, Han Z, et al. TCR-T immunotherapy: The challenges and solutions. Frontiers in Oncology. 2021;**11**:794183

[3] Takahashi K, Yamanaka S. Induction of pluripotent stem cells from mouse embryonic and adult fibroblast cultures by defined factors. Cell. 2006;**126**(4):663-676

[4] Yu J, Vodyanik MA, Smuga-Otto K, Antosiewicz-Bourget J, Frane JL, Tian S, et al. Induced pluripotent stem cell lines derived from human somatic cells. Science (New York, N.Y.). 2007;**318**(5858):1917-1920

[5] Ghosh Z, Wilson KD, Wu Y, Hu S, Quertermous T, Wu JC. Persistent donor cell gene expression among human induced pluripotent stem cells contributes to differences with human embryonic stem cells. PLoS One. 2010;**5**(2):e8975

[6] Marchetto MC, Yeo GW, Kainohana O, Marsala M, Gage FH, Muotri AR. Transcriptional signature and memory retention of human-induced pluripotent stem cells. PLoS One. 2009;**4**(9):e7076

[7] Kennedy M, Awong G, Sturgeon CM, Ditadi A, LaMotte-Mohs R, Zúñiga-Pflücker JC, et al. T lymphocyte potential marks the emergence of definitive hematopoietic progenitors in human pluripotent stem cell differentiation cultures. Cell Reports. 2012;**2**(6):1722-1735

[8] Qasim W, Zhan H, Samarasinghe S, Adams S, Amrolia P, Stafford S, et al. Molecular remission of infant B-ALL after infusion of universal TALEN gene-edited CAR T cells. Science Translational Medicine. 2017;**9**(374):eaaj2013

[9] Xue D, Lu S, Zhang H, Zhang L, Dai Z, Kaufman DS, et al. Induced pluripotent stem cell-derived engineered T cells, natural killer cells, macrophages, and dendritic cells in immunotherapy. Trends in Biotechnology. 2023;**41**(7):907-922

[10] Shortman K, Wu L. Early T lymphocyte progenitors. Annual Review of Immunology. 1996;**14**:29-47

[11] Rosenberg SA, Restifo NP. Adoptive cell transfer as personalized immunotherapy for human cancer. Science (New York, N.Y.). 2015;**348**(6230):62-68

[12] June CH, O'Connor RS, Kawalekar OU, Ghassemi S, Milone MC. CAR T cell immunotherapy for human cancer. Science (New York, N.Y.). 2018;**359**(6382):1361-1365

[13] Tombácz I, Laczkó D, Shahnawaz H, Muramatsu H, Natesan A, Yadegari A, et al. Highly efficient CD4+ T cell targeting and genetic recombination using engineered CD4+ cell-homing mRNA-LNPs. Molecular Therapy: The Journal of the American Society of Gene Therapy. 2021;**29**(11):3293-3304

[14] Chen J, Ye Z, Huang C, Qiu M, Song D, Li Y, et al. Lipid nanoparticle-mediated lymph node-targeting delivery of mRNA cancer vaccine elicits robust

CD8(+) T cell response. Proceedings of the National Academy of Sciences of the United States of America. 2022;**119**(34):e2207841119

[15] Maeda T, Nagano S, Kashima S, Terada K, Agata Y, Ichise H, et al. Regeneration of tumor-antigen-specific cytotoxic T lymphocytes from iPSCs transduced with exogenous TCR genes. Molecular Therapy Methods & Clinical Development. 2020;**19**:250-260

[16] Kawamoto H, Masuda K, Nagano S. Regeneration of antigen-specific T cells by using induced pluripotent stem cell (iPSC) technology. International Immunology. 2021;**33**(12):827-833

[17] Hochedlinger K, Jaenisch R. Monoclonal mice generated by nuclear transfer from mature B and T donor cells. Nature. 2002;**415**(6875):1035-1038

[18] Nagano S, Maeda T, Ichise H, Kashima S, Ohtaka M, Nakanishi M, et al. High frequency production of T cell-derived iPSC clones capable of generating potent cytotoxic T cells. Molecular Therapy Methods & Clinical Development. 2020;**16**:126-135

[19] Kamaldinov T, Zhang L, Wang L, Ye Z. Generation of induced pluripotent stem cell lines from helper and cytotoxic T cells of healthy individuals. Stem Cell Research. 2023;**69**:103113

[20] Flippe L, Gaignerie A, Sérazin C, Baron O, Saulquin X, Anegon I, et al. Generation of CD34(+)CD43(+) hematopoietic progenitors to induce thymocytes from human pluripotent stem cells. Cell. 2022;**11**(24):4046

[21] PratumkaewP,LuanpitpongS,KlaihmonP, Lorthongpanich C, Laowtammathron C, Meesa S, et al. Episomal vector-based generation of human induced pluripotent stem cell line MUSIi020-a from peripheral blood T-cells. Stem Cell Research. 2022;**64**:102929

[22] Maeda T, Nagano S, Ichise H, Kataoka K, Yamada D, Ogawa S, et al. Regeneration of CD8αβ T cells from T-cell-derived iPSC imparts potent tumor antigen-specific cytotoxicity. Cancer Research. 2016;**76**(23):6839-6850

[23] Vizcardo R, Masuda K, Yamada D, Ikawa T, Shimizu K, Fujii S, et al. Regeneration of human tumor antigen-specific T cells from iPSCs derived from mature CD8(+) T cells. Cell Stem Cell. 2013;**12**(1):31-36

[24] Nishimura T, Kaneko S, Kawana-Tachikawa A, Tajima Y, Goto H, Zhu D, et al. Generation of rejuvenated antigen-specific T cells by reprogramming to pluripotency and redifferentiation. Cell Stem Cell. 2013;**12**(1):114-126

[25] Minagawa A, Yoshikawa T, Yasukawa M, Hotta A, Kunitomo M, Iriguchi S, et al. Enhancing T cell receptor stability in rejuvenated iPSC-derived T cells improves their use in cancer immunotherapy. Cell Stem Cell. 2018;**23**(6):850-858.e854

[26] Themeli M, Kloss CC, Ciriello G, Fedorov VD, Perna F, Gonen M, et al. Generation of tumor-targeted human T lymphocytes from induced pluripotent stem cells for cancer therapy. Nature Biotechnology. 2013;**31**(10):928-933

[27] Iriguchi S, Yasui Y, Kawai Y, Arima S, Kunitomo M, Sato T, et al. A clinically applicable and scalable method to regenerate T-cells from iPSCs for off-the-shelf T-cell immunotherapy. Nature Communications. 2021;**12**(1):430

[28] Jing R, Scarfo I, Najia MA, Lummertz da Rocha E, Han A, Sanborn M, et al. EZH1 repression generates mature

iPSC-derived CAR T cells with enhanced antitumor activity. Cell Stem Cell. 2022;**29**(8):1181-1196.e1186

[29] Montel-Hagen A, Seet CS, Li S, Chick B, Zhu Y, Chang P, et al. Organoid-induced differentiation of conventional T cells from human pluripotent stem cells. Cell Stem Cell. 2019;**24**(3):376-389.e378

[30] Wang Z, McWilliams-Koeppen HP, Reza H, Ostberg JR, Chen W, Wang X, et al. 3D-organoid culture supports differentiation of human CAR(+) iPSCs into highly functional CAR T cells. Cell Stem Cell. 2022;**29**(4):515-527.e518

[31] Clements WK, Kim AD, Ong KG, Moore JC, Lawson ND, Traver D. A somitic Wnt16/notch pathway specifies haematopoietic stem cells. Nature. 2011;**474**(7350):220-224

[32] Hadland BK, Huppert SS, Kanungo J, Xue Y, Jiang R, Gridley T, et al. A requirement for Notch1 distinguishes 2 phases of definitive hematopoiesis during development. Blood. 2004;**104**(10):3097-3105

[33] Kumano K, Chiba S, Kunisato A, Sata M, Saito T, Nakagami-Yamaguchi E, et al. Notch1 but not Notch2 is essential for generating hematopoietic stem cells from endothelial cells. Immunity. 2003;**18**(5):699-711

[34] Ditadi A, Sturgeon CM, Tober J, Awong G, Kennedy M, Yzaguirre AD, et al. Human definitive haemogenic endothelium and arterial vascular endothelium represent distinct lineages. Nature Cell Biology. 2015;**17**(5):580-591

[35] Leung A, Zulick E, Skvir N, Vanuytsel K, Morrison TA, Naing ZH, et al. Notch and aryl hydrocarbon receptor signaling impact definitive hematopoiesis from human pluripotent stem cells. Stem Cells (Dayton, Ohio). 2018;**36**(7):1004-1019

[36] Heinze D, Park S, McCracken A, Haratianfar M, Lindstrom J, Villacorta-Martin C, et al. Notch activation during early mesoderm induction modulates emergence of the T/NK cell lineage from human iPSCs. Stem Cell Reports. 2022;**17**(12):2610-2628

[37] Good ML, Vizcardo R, Maeda T, Tamaoki N, Malekzadeh P, Kawamoto H, et al. Using human induced pluripotent stem cells for the generation of tumor antigen-specific T cells. Journal of Visualized Experiments: JoVE. 2019;**152**:e59997

[38] Trotman-Grant AC, Mohtashami M, De Sousa CJ, Martinez EC, Lee D, Teichman S, et al. DL4-μbeads induce T cell lineage differentiation from stem cells in a stromal cell-free system. Nature Communications. 2021;**12**(1):5023

[39] Seet CS, He C, Bethune MT, Li S, Chick B, Gschweng EH, et al. Generation of mature T cells from human hematopoietic stem and progenitor cells in artificial thymic organoids. Nature Methods. 2017;**14**(5):521-530

[40] O'Reilly RJ, Prockop S, Hasan AN, Koehne G, Doubrovina E. Virus-specific T-cell banks for 'off the shelf' adoptive therapy of refractory infections. Bone Marrow Transplantation. 2016;**51**(9):1163-1172

[41] Torikai H, Reik A, Liu PQ, Zhou Y, Zhang L, Maiti S, et al. A foundation for universal T-cell based immunotherapy: T cells engineered to express a CD19-specific chimeric-antigen-receptor and eliminate expression of endogenous TCR. Blood. 2012;**119**(24):5697-5705

[42] Yang D, Ling X, Liu F, Duan J, Bai F, Hu A. Generation of a human induced pluripotent stem cell line (SYSUTFi001-a) from infiltrating cytotoxic T cells in hepatocellular

carcinoma (HCC). Stem Cell Research. 2022;**65**:102962

[43] Itoh M, Kawagoe S, Nakagawa H, Asahina A, Okano HJ. Generation of induced pluripotent stem cell (iPSC) from NY-ESO-I-specific cytotoxic T cells isolated from the melanoma patient with minor HLAs: The practical pilot study for the adoptive immunotherapy for melanoma using iPSC technology. Experimental Dermatology. 2023;**32**(2):126-134

[44] Timmermans F, Velghe I, Vanwalleghem L, De Smedt M, Van Coppernolle S, Taghon T, et al. Generation of T cells from human embryonic stem cell-derived hematopoietic zones. Journal of Immunology (Baltimore, Md: 1950). 2009;**182**(11):6879-6888

[45] Wang B, Iriguchi S, Waseda M, Ueda N, Ueda T, Xu H, et al. Generation of hypoimmunogenic T cells from genetically engineered allogeneic human induced pluripotent stem cells. Nature Biomedical Engineering. 2021;**5**(5):429-440

[46] Johnson LA, Morgan RA, Dudley ME, Cassard L, Yang JC, Hughes MS, et al. Gene therapy with human and mouse T-cell receptors mediates cancer regression and targets normal tissues expressing cognate antigen. Blood. 2009;**114**(3):535-546

[47] Robbins PF, Kassim SH, Tran TL, Crystal JS, Morgan RA, Feldman SA, et al. A pilot trial using lymphocytes genetically engineered with an NY-ESO-1-reactive T-cell receptor: Long-term follow-up and correlates with response. Clinical Cancer Research: An Official Journal of the American Association for Cancer Research. 2015;**21**(5):1019-1027

[48] Harada S, Ando M, Ando J, Ishii M, Yamaguchi T, Yamazaki S, et al. Dual-antigen targeted iPSC-derived chimeric antigen receptor-T cell therapy for refractory lymphoma. Molecular Therapy: The Journal of the American Society of Gene Therapy. 2022;**30**(2):534-549

[49] Ando M, Ando J, Yamazaki S, Ishii M, Sakiyama Y, Harada S, et al. Long-term eradication of extranodal natural killer/T-cell lymphoma, nasal type, by induced pluripotent stem cell-derived Epstein-Barr virus-specific rejuvenated T cells in vivo. Haematologica. 2020;**105**(3):796-807

[50] Rooney CM, Smith CA, Ng CY, Loftin S, Li C, Krance RA, et al. Use of gene-modified virus-specific T lymphocytes to control Epstein-Barr-virus-related lymphoproliferation. Lancet (London, England). 1995;**345**(8941):9-13

[51] Comoli P, Pedrazzoli P, Maccario R, Basso S, Carminati O, Labirio M, et al. Cell therapy of stage IV nasopharyngeal carcinoma with autologous Epstein-Barr virus-targeted cytotoxic T lymphocytes. Journal of Clinical Oncology: Official Journal of the American Society of Clinical Oncology. 2005;**23**(35):8942-8949

[52] Gottschalk S, Edwards OL, Sili U, Huls MH, Goltsova T, Davis AR, et al. Generating CTLs against the subdominant Epstein-Barr virus LMP1 antigen for the adoptive immunotherapy of EBV-associated malignancies. Blood. 2003;**101**(5):1905-1912

[53] Heslop HE, Slobod KS, Pule MA, Hale GA, Rousseau A, Smith CA, et al. Long-term outcome of EBV-specific T-cell infusions to prevent or treat EBV-related lymphoproliferative disease in transplant recipients. Blood. 2010;**115**(5):925-935

[54] Bollard CM, Gottschalk S, Leen AM, Weiss H, Straathof KC, Carrum G,

et al. Complete responses of relapsed lymphoma following genetic modification of tumor-antigen presenting cells and T-lymphocyte transfer. Blood. 2007;**110**(8):2838-2845

[55] Bollard CM, Rooney CM, Heslop HE. T-cell therapy in the treatment of post-transplant lymphoproliferative disease. Nature Reviews Clinical Oncology. 2012;**9**(9):510-519

[56] Honda T, Ando M, Ando J, Ishii M, Sakiyama Y, Ohara K, et al. Sustainable tumor-suppressive effect of iPSC-derived rejuvenated T cells targeting cervical cancers. Molecular Therapy: The Journal of the American Society of Gene Therapy. 2020;**28**(11):2394-2405

[57] Kashima S, Maeda T, Masuda K, Nagano S, Inoue T, Takeda M, et al. Cytotoxic T lymphocytes regenerated from iPS cells have therapeutic efficacy in a patient-derived xenograft solid tumor model. iScience. 2020;**23**(4):100998

[58] Xu W, Cherrier DE, Chea S, Vosshenrich C, Serafini N, Petit M, et al. An Id2(RFP)-reporter mouse redefines innate lymphoid cell precursor potentials. Immunity. 2019;**50**(4):1054-1068.e1053

[59] Mehta RS, Randolph B, Daher M, Rezvani K. NK cell therapy for hematologic malignancies. International Journal of Hematology. 2018;**107**(3):262-270

[60] Knorr DA, Bachanova V, Verneris MR, Miller JS. Clinical utility of natural killer cells in cancer therapy and transplantation. Seminars in Immunology. 2014;**26**(2):161-172

[61] Ni Z, Knorr DA, Clouser CL, Hexum MK, Southern P, Mansky LM, et al. Human pluripotent stem cells produce natural killer cells that mediate anti-HIV-1 activity by utilizing diverse cellular mechanisms. Journal of Virology. 2011;**85**(1):43-50

[62] Bock AM, Knorr D, Kaufman DS. Development, expansion, and in vivo monitoring of human NK cells from human embryonic stem cells (hESCs) and and induced pluripotent stem cells (iPSCs). Journal of Visualized Experiments: JoVE. 2013;**74**:e50337

[63] Knorr DA, Ni Z, Hermanson D, Hexum MK, Bendzick L, Cooper LJ, et al. Clinical-scale derivation of natural killer cells from human pluripotent stem cells for cancer therapy. Stem Cells Translational Medicine. 2013;**2**(4):274-283

[64] Zeng J, Tang SY, Toh LL, Wang S. Generation of "off-the-shelf" natural killer cells from peripheral blood cell-derived induced pluripotent stem cells. Stem Cell Reports. 2017;**9**(6):1796-1812

[65] Cichocki F, Bjordahl R, Gaidarova S, Mahmood S, Abujarour R, Wang H, et al. iPSC-derived NK cells maintain high cytotoxicity and enhance in vivo tumor control in concert with T cells and anti-PD-1 therapy. Science Translational Medicine. 2020;**12**(568):eaaz5618

[66] Hermanson DL, Bendzick L, Pribyl L, McCullar V, Vogel RI, Miller JS, et al. Induced pluripotent stem cell-derived natural killer cells for treatment of ovarian cancer. Stem Cells (Dayton, Ohio). 2016;**34**(1):93-101

[67] Euchner J, Sprissler J, Cathomen T, Fürst D, Schrezenmeier H, Debatin KM, et al. Natural killer cells generated from human induced pluripotent stem cells mature to CD56(bright)CD16(+) NKp80(+/−)In-vitro and express KIR2DL2/DL3 and KIR3DL1. Frontiers in Immunology. 2021;**12**:640672

[68] Ni Z, Knorr DA, Bendzick L, Allred J, Kaufman DS. Expression of chimeric receptor CD4ζ by natural killer cells derived from human pluripotent stem cells improves in vitro activity but does not enhance suppression of HIV infection in vivo. Stem Cells (Dayton, Ohio). 2014;**32**(4):1021-1031

[69] Lupo KB, Moon JI, Chambers AM, Matosevic S. Differentiation of natural killer cells from induced pluripotent stem cells under defined, serum- and feeder-free conditions. Cytotherapy. 2021;**23**(10):939-952

[70] Daher M, Melo Garcia L, Li Y, Rezvani K. CAR-NK cells: The next wave of cellular therapy for cancer. Clinical & Translational Immunology. 2021;**10**(4):e1274

[71] Boissel L, Betancur-Boissel M, Lu W, Krause DS, Van Etten RA, Wels WS, et al. Retargeting NK-92 cells by means of CD19- and CD20-specific chimeric antigen receptors compares favorably with antibody-dependent cellular cytotoxicity. Oncoimmunology. 2013;**2**(10):e26527

[72] Bagheri Y, Barati A, Aghebati-Maleki A, Aghebati-Maleki L, Yousefi M. Current progress in cancer immunotherapy based on natural killer cells. Cell Biology International. 2021;**45**(1):2-17

[73] Hsu LJ, Liu CL, Kuo ML, Shen CN, Shen CR. An alternative cell therapy for cancers: Induced pluripotent stem cell (iPSC)-derived natural killer cells. Biomedicine. 2021;**9**(10):1323

[74] Ruggeri L, Capanni M, Urbani E, Perruccio K, Shlomchik WD, Tosti A, et al. Effectiveness of donor natural killer cell alloreactivity in mismatched hematopoietic transplants. Science (New York, N.Y.). 2002;**295**(5562):2097-2100

[75] Heipertz EL, Zynda ER, Stav-Noraas TE, Hungler AD, Boucher SE, Kaur N, et al. Current perspectives on "off-the-shelf" allogeneic NK and CAR-NK cell therapies. Frontiers in Immunology. 2021;**12**:732135

[76] Lu H, Zhao X, Li Z, Hu Y, Wang H. From CAR-T cells to CAR-NK cells: A developing immunotherapy method for hematological malignancies. Frontiers in Oncology. 2021;**11**:720501

[77] Xie G, Dong H, Liang Y, Ham JD, Rizwan R, Chen J. CAR-NK cells: A promising cellular immunotherapy for cancer. eBioMedicine. 2020;**59**:102975

[78] Li Y, Hermanson DL, Moriarity BS, Kaufman DS. Human iPSC-derived natural killer cells engineered with chimeric antigen receptors enhance anti-tumor activity. Cell Stem Cell. 2018;**23**(2):181-192.e185

[79] Mrózek E, Anderson P, Caligiuri MA. Role of interleukin-15 in the development of human CD56+ natural killer cells from CD34+ hematopoietic progenitor cells. Blood. 1996;**87**(7):2632-2640

[80] Conlon KC, Lugli E, Welles HC, Rosenberg SA, Fojo AT, Morris JC, et al. Redistribution, hyperproliferation, activation of natural killer cells and CD8 T cells, and cytokine production during first-in-human clinical trial of recombinant human interleukin-15 in patients with cancer. Journal of Clinical Oncology: Official Journal of the American Society of Clinical Oncology. 2015;**33**(1):74-82

[81] Woan KV, Kim H, Bjordahl R, Davis ZB, Gaidarova S, Goulding J, et al. Harnessing features of adaptive NK cells to generate iPSC-derived NK cells for enhanced immunotherapy. Cell Stem Cell. 2021;**28**(12):2062-2075.e2065

[82] Zhu H, Blum RH, Bernareggi D, Ask EH, Wu Z, Hoel HJ, et al. Metabolic reprograming via deletion of CISH in human iPSC-derived NK cells promotes In vivo persistence and enhances anti-tumor activity. Cell Stem Cell. 2020;**27**(2):224-237.e226

[83] Ma S, Caligiuri MA, Yu J. Harnessing IL-15 signaling to potentiate NK cell-mediated cancer immunotherapy. Trends in Immunology. 2022;**43**(10):833-847

[84] Chiu E, Felices M, Cichocki F, Davis Z, Wang H, Tuninga K, et al. Anti-NKG2C/IL-15/anti-CD33 killer engager directs primary and iPSC-derived NKG2C(+) NK cells to target myeloid leukemia. Molecular Therapy: The Journal of the American Society of Gene Therapy. 2021;**29**(12):3410-3421

[85] Snyder KM, Hullsiek R, Mishra HK, Mendez DC, Li Y, Rogich A, et al. Expression of a recombinant high affinity IgG fc receptor by engineered NK cells as a docking platform for therapeutic mAbs to target cancer cells. Frontiers in Immunology. 2018;**9**:2873

[86] Zhu H, Blum RH, Bjordahl R, Gaidarova S, Rogers P, Lee TT, et al. Pluripotent stem cell-derived NK cells with high-affinity noncleavable CD16a mediate improved antitumor activity. Blood. 2020;**135**(6):399-410

[87] Meng F, Zhang S, Xie J, Zhou Y, Wu Q, Lu B, et al. Leveraging CD16 fusion receptors to remodel the immune response for enhancing anti-tumor immunotherapy in iPSC-derived NK cells. Journal of Hematology & Oncology. 2023;**16**(1):62

[88] Chakrabarty K, Shetty R, Argulwar S, Das D, Ghosh A. Induced pluripotent stem cell-based disease modeling and prospective immune therapy for coronavirus disease 2019. Cytotherapy. 2022;**24**(3):235-248

[89] Zhang L, Tian L, Dai X, Yu H, Wang J, Lei A, et al. Pluripotent stem cell-derived CAR-macrophage cells with antigen-dependent anti-cancer cell functions. Journal of Hematology & Oncology. 2020;**13**(1):153

[90] Kambal A, Mitchell G, Cary W, Gruenloh W, Jung Y, Kalomoiris S, et al. Generation of HIV-1 resistant and functional macrophages from hematopoietic stem cell-derived induced pluripotent stem cells. Molecular Therapy: The Journal of the American Society of Gene Therapy. 2011;**19**(3):584-593

[91] Aflaki E, Stubblefield BK, Maniwang E, Lopez G, Moaven N, Goldin E, et al. Macrophage models of Gaucher disease for evaluating disease pathogenesis and candidate drugs. Science Translational Medicine. 2014;**6**(240):240ra273

[92] Choi KD, Vodyanik M, Slukvin II. Hematopoietic differentiation and production of mature myeloid cells from human pluripotent stem cells. Nature Protocols. 2011;**6**(3):296-313

[93] Lachmann N, Happle C, Ackermann M, Lüttge D, Wetzke M, Merkert S, et al. Gene correction of human induced pluripotent stem cells repairs the cellular phenotype in pulmonary alveolar proteinosis. American Journal of Respiratory and Critical Care Medicine. 2014;**189**(2):167-182

[94] Takamatsu K, Ikeda T, Haruta M, Matsumura K, Ogi Y, Nakagata N, et al. Degradation of amyloid beta by human induced pluripotent stem cell-derived macrophages expressing Neprilysin-2. Stem Cell Research. 2014;**13**(3 Pt A): 442-453

[95] Zhang H, Xue C, Shah R, Bermingham K, Hinkle CC, Li W, et al. Functional analysis and transcriptomic profiling of iPSC-derived macrophages and their application in modeling Mendelian disease. Circulation Research. 2015;**117**(1):17-28

[96] Gutbier S, Wanke F, Dahm N, Rümmelin A, Zimmermann S, Christensen K, et al. Large-scale production of human iPSC-derived macrophages for drug screening. International Journal of Molecular Sciences. 2020;**21**(13):4808

[97] Gutbier S, Wanke F, Dahm N, Rümmelin A, Zimmermann S, Christensen K, et al. Large-scale production of human iPSC-derived macrophages for drug screening. International Journal of Molecular Sciences. 2020;**21**(13):4808

[98] Lopez-Yrigoyen M, Fidanza A, Cassetta L, Axton RA, Taylor AH, Meseguer-Ripolles J, et al. A human iPSC line capable of differentiating into functional macrophages expressing ZsGreen: A tool for the study and in vivo tracking of therapeutic cells. Philosophical Transactions of the Royal Society of London Series B, Biological Sciences. 2018;**373**(1750):20170219

[99] Lopez-Yrigoyen M, Fidanza A, Cassetta L, Axton RA, Taylor AH, Meseguer-Ripolles J, et al. A human iPSC line capable of differentiating into functional macrophages expressing ZsGreen: A tool for the study and in vivo tracking of therapeutic cells. Philosophical Transactions of the Royal Society of London Series B, Biological Sciences. 2018;**373**(1750):20170219

[100] Cao X, Yakala GK, van den Hil FE, Cochrane A, Mummery CL, Orlova VV. Differentiation and functional comparison of monocytes and macrophages from hiPSCs with peripheral blood derivatives. Stem Cell Reports. 2019;**12**(6):1282-1297

[101] Cui D, Franz A, Fillon SA, Jannetti L, Isambert T, Fundel-Clemens K, et al. High-yield human induced pluripotent stem cell-derived monocytes and macrophages are functionally comparable with primary cells. Frontiers in Cell and Developmental Biology. 2021;**9**:656867

[102] Klichinsky M, Ruella M, Shestova O, Lu XM, Best A, Zeeman M, et al. Human chimeric antigen receptor macrophages for cancer immunotherapy. Nature Biotechnology. 2020;**38**(8):947-953

[103] Senju S, Haruta M, Matsumura K, Matsunaga Y, Fukushima S, Ikeda T, et al. Generation of dendritic cells and macrophages from human induced pluripotent stem cells aiming at cell therapy. Gene Therapy. 2011;**18**(9):874-883

[104] Miyashita A, Fukushima S, Nakahara S, Kubo Y, Tokuzumi A, Yamashita J, et al. Immunotherapy against metastatic melanoma with human iPS cell-derived myeloid cell lines producing type I interferons. Cancer Immunology Research. 2016;**4**(3):248-258

[105] Kitadani J, Ojima T, Iwamoto H, Tabata H, Nakamori M, Nakamura M, et al. Cancer vaccine therapy using carcinoembryonic antigen—Expressing dendritic cells generated from induced pluripotent stem cells. Scientific Reports. 2018;**8**(1):4569

[106] Senju S, Haruta M, Matsunaga Y, Fukushima S, Ikeda T, Takahashi K, et al. Characterization of dendritic cells and macrophages generated by directed differentiation from mouse induced pluripotent stem cells. Stem Cells (Dayton, Ohio). 2009;**27**(5):1021-1031

[107] Zhang Q, Fujino M, Iwasaki S, Hirano H, Cai S, Kitajima Y, et al.

Generation and characterization of regulatory dendritic cells derived from murine induced pluripotent stem cells. Scientific Reports. 2014;**4**:3979

[108] Rosenberg SA, Yang JC, Restifo NP. Cancer immunotherapy: Moving beyond current vaccines. Nature Medicine. 2004;**10**(9):909-915

[109] Fukushima S, Miyashita A, Kuriyama H, Kimura T, Mizuhashi S, Kubo Y, et al. Future prospects for cancer immunotherapy using induced pluripotent stem cell-derived dendritic cells or macrophages. Experimental Dermatology. 2023;**32**(3):290-296

[110] Tominaga S, Ojima T, Miyazawa M, Iwamoto H, Kitadani J, Maruoka S, et al. Induced pluripotent stem cell-derived dendritic cell vaccine therapy genetically modified on the ubiquitin-proteasome system. Gene Therapy. 2023;**30**(7-8):552-559

[111] Takahashi K, Tanabe K, Ohnuki M, Narita M, Ichisaka T, Tomoda K, et al. Induction of pluripotent stem cells from adult human fibroblasts by defined factors. Cell. 2007;**131**(5):861-872

[112] Lowry WE, Richter L, Yachechko R, Pyle AD, Tchieu J, Sridharan R, et al. Generation of human induced pluripotent stem cells from dermal fibroblasts. Proceedings of the National Academy of Sciences of the United States of America. 2008;**105**(8):2883-2888

[113] Ng ES, Davis R, Stanley EG, Elefanty AG. A protocol describing the use of a recombinant protein-based, animal product-free medium (APEL) for human embryonic stem cell differentiation as spin embryoid bodies. Nature Protocols. 2008;**3**(5):768-776

[114] Mukherjee C, Hale C, Mukhopadhyay S. A simple multistep protocol for differentiating human induced pluripotent stem cells into functional macrophages. Methods in Molecular Biology (Clifton, NJ). 2018;**1784**:13-28

[115] Shi J, Xue C, Liu W, Zhang H. Differentiation of human-induced pluripotent stem cells to macrophages for disease modeling and functional genomics. Current Protocols in Stem Cell Biology. 2019;**48**(1):e74

[116] Vaughan-Jackson A, Stodolak S, Ebrahimi KH, Browne C, Reardon PK, Pires E, et al. Differentiation of human induced pluripotent stem cells to authentic macrophages using a defined, serum-free, open-source medium. Stem Cell Reports. 2021;**16**(7):1735-1748

[117] Lee AS, Tang C, Rao MS, Weissman IL, Wu JC. Tumorigenicity as a clinical hurdle for pluripotent stem cell therapies. Nature Medicine. 2013;**19**(8):998-1004

[118] Nori S, Okada Y, Nishimura S, Sasaki T, Itakura G, Kobayashi Y, et al. Long-term safety issues of iPSC-based cell therapy in a spinal cord injury model: Oncogenic transformation with epithelial-mesenchymal transition. Stem Cell Reports. 2015;**4**(3):360-373

[119] Huang CY, Liu CL, Ting CY, Chiu YT, Cheng YC, Nicholson MW, et al. Human iPSC banking: Barriers and opportunities. Journal of Biomedical Science. 2019;**26**(1):87

[120] Tang C, Lee AS, Volkmer JP, Sahoo D, Nag D, Mosley AR, et al. An antibody against SSEA-5 glycan on human pluripotent stem cells enables removal of teratoma-forming cells. Nature Biotechnology. 2011;**29**(9):829-834

[121] Choo AB, Tan HL, Ang SN, Fong WJ, Chin A, Lo J, et al. Selection against undifferentiated human

embryonic stem cells by a cytotoxic antibody recognizing podocalyxin-like protein-1. Stem Cells (Dayton, Ohio). 2008;**26**(6):1454-1463

[122] Ben-David U, Gan QF, Golan-Lev T, Arora P, Yanuka O, Oren YS, et al. Selective elimination of human pluripotent stem cells by an oleate synthesis inhibitor discovered in a high-throughput screen. Cell Stem Cell. 2013;**12**(2):167-179

[123] Lee MO, Moon SH, Jeong HC, Yi JY, Lee TH, Shim SH, et al. Inhibition of pluripotent stem cell-derived teratoma formation by small molecules. Proceedings of the National Academy of Sciences of the United States of America. 2013;**110**(35):E3281-E3290

[124] Ciceri F, Bonini C, Gallo-Stampino C, Bordignon C. Modulation of GvHD by suicide-gene transduced donor T lymphocytes: Clinical applications in mismatched transplantation. Cytotherapy. 2005;**7**(2):144-149

[125] Ciceri F, Bonini C, Marktel S, Zappone E, Servida P, Bernardi M, et al. Antitumor effects of HSV-TK-engineered donor lymphocytes after allogeneic stem-cell transplantation. Blood. 2007;**109**(11):4698-4707

[126] Ando M, Nishimura T, Yamazaki S, Yamaguchi T, Kawana-Tachikawa A, Hayama T, et al. A safeguard system for induced pluripotent stem cell-derived rejuvenated T cell therapy. Stem Cell Reports. 2015;**5**(4):597-608

[127] Pozo MR, Meredith GW, Entcheva E. Human iPSC-Cardiomyocytes as an experimental model to study epigenetic modifiers of electrophysiology. Cell. 2022;**11**(2):200

[128] Pozo MR, Meredith GW. Entcheva E: Human iPSC-Cardiomyocytes as an experimental model to study epigenetic modifiers of electrophysiology. Cell. 2022;**11**(2):200

[129] Tanaka T, Takahashi K, Yamane M, Tomida S, Nakamura S, Oshima K, et al. Induced pluripotent stem cells from CINCA syndrome patients as a model for dissecting somatic mosaicism and drug discovery. Blood. 2012;**120**(6):1299-1308

[130] Hirata S, Senju S, Matsuyoshi H, Fukuma D, Uemura Y, Nishimura Y. Prevention of experimental autoimmune encephalomyelitis by transfer of embryonic stem cell-derived dendritic cells expressing myelin oligodendrocyte glycoprotein peptide along with TRAIL or programmed death-1 ligand. Journal of Immunology (Baltimore, Md: 1950). 2005;**174**(4):1888-1897

[131] Hirata S, Matsuyoshi H, Fukuma D, Kurisaki A, Uemura Y, Nishimura Y, et al. Involvement of regulatory T cells in the experimental autoimmune encephalomyelitis-preventive effect of dendritic cells expressing myelin oligodendrocyte glycoprotein plus TRAIL. Journal of Immunology (Baltimore, Md: 1950). 2007;**178**(2):918-925

[132] Ikeda T, Hirata S, Fukushima S, Matsunaga Y, Ito T, Uchino M, et al. Dual effects of TRAIL in suppression of autoimmunity: The inhibition of Th1 cells and the promotion of regulatory T cells. Journal of Immunology (Baltimore, Md: 1950). 2010;**185**(9):5259-5267

[133] Ikeda T, Hirata S, Takamatsu K, Haruta M, Tsukamoto H, Ito T, et al. Suppression of Th1-mediated autoimmunity by embryonic stem cell-derived dendritic cells. PLoS One. 2014;**9**(12):e115198

Section 3

Human Pluripotent Stem Cells for Tissue Engineering

Chapter 6

Advanced Hydrogel for Physiological 3D Colonies of Pluripotent Stem Cells

Quan Li, Guangyan Qi and Xiuzhi Susan Sun

Abstract

Human induced pluripotent stem cells (hiPSCs) demonstrated great potential in basic research, disease modeling, drug development, cell therapeutics, and regenerative medicine, as various distinct somatic cell types such as hepatocytes can be derived from hiPSCs. However, highly efficient hiPSC to somatic cell differentiation has not yet been achieved because of various challenging problems, one of which is less-optimal culture methods for hiPSC expansion. Conventionally, hiPSCs have been cultured as monolayers on flat surfaces, usually resulting in unstable genetic integrity, reduced pluripotency, and spontaneous differentiation after numerous passages. Recently, three-dimensional (3D) spheroids of hiPSCs have shown potential for somatic cell differentiations. However, these hiPSC spheroids are generated using 2D-cultured cells in either nonadherent U-bottom 96-well plates or agarose microarray molding plates, in which single hiPSCs are forced to aggregate into spheroids. These "aggregation molding" methods are neither typically suited for large-scale hiPSC manufacturing nor for tissue engineering. In addition, the aggregated hiPSC spheroids present limited functions compared to physiologically formed hiPSC 3D colonies. In this chapter, advanced 3D cell culture technologies will be reviewed, and comprehensive discussions and future development will be provided and suggested.

Keywords: hiPSC, 3D culture, PGmatrix, hydrogel, peptide, 3D bioprinting

1. Introduction

Human pluripotent stem cells (hPSCs), including human embryonic stem cells (hESCs) and human induced pluripotent stem cells (hiPSCs), are able to self-renew indefinitely in theory and to differentiate into almost all somatic cell types; thus, they have drawn great attention in the research community and hold great potential to improve human health [1]. Unlike hESCs that are obtained from early-stage embryos, hiPSCs can be generated from somatic cells through ectopic expression of defined transcription factors [2] and therefore avoid ethical problems related to the use of hESCs. In addition, the successful generation and specific differentiation of patient-specific hiPSCs provide new approaches for disease modeling, drug screening/toxicity testing as well as personalized cell therapies. However, these applications have been hindered in part by the current less-optimal culture technologies that may reduce or mask some of the concerns related with tumorigenesis, low viability/retention, and uncontrolled *in vivo* differentiation. Commonly used culture methods cannot completely fulfill the demand for hiPSCs in view of genetic quality, growth performance, stemness, functionality, and differentiation potentiality [3].

Recently, three-dimensional (3D) spheroids of pluripotent stem cells (PSCs), including hiPSCs, have been demonstrated to have great potential for differentiation into various types of organoids, including hepatic cells [4, 5], lung organoids [6], as well as hair-bearing human skin [7]. In most organoid formation studies, the initial spheroids were produced by forced aggregation of single stem cells in suspension/nonadherent plate or agarose microarray molding plates [5, 7, 8]. Despite the advantages observed with 3D differentiation, the starting PSCs were still cultured in 2D, which is not suited for large-scale manufacturing [5, 7]. In this chapter, we will review advanced developments in hiPSC culture methods with emphasis on 3D culture and discuss the benefits and future developments.

2. Culture methods for production of high-quality hiPSC spheroids

2.1 From 2D to 3D cultures

Traditionally, hiPSCs are seeded onto a thin layer of substrate or feeder and grow as colonies of monolayer. Mitotically inactivated mouse embryonic fibroblast (MEF) [9] and Matrigel extracted from mouse sarcoma [10] are the most commonly used feeder layer and substrates for hiPSC culture. For clinical translation, creating a stable, scalable, more defined, and cost-effective culture environment for hiPSC is of high importance. In the past decades, researchers have put tremendous efforts into seeking alternative, well-defined xeno-free matrices, such as laminin [11–13], E-cadherin [14, 15], fibronectin [16], vitronectin [17], synthetic polymers [18–21], and synthetic peptides [22–25]. However, most of these matrices are limited to 2D culture, in which a monolayer of hiPSCs grows on top of matrix-coated flat or bead surfaces. Obviously, these culture systems cannot support large-scale hiPSC production as cell proliferation and expansion requires a large surface area. More importantly, 2D culture is not representing the *in vivo* physiological environment. The side-to-side cell contact lack of appropriate stem cell niche often leads to unwanted gene expression, resulting in inefficiency of targeted differentiation [26].

Studies on mesenchymal stem cells (MSCs) have shown that by adopting 3D configuration, pluripotency of MSCs was promoted [27, 28]. Similar aggregation techniques were also used before initiating differentiation [5, 7]. However, such suspension methods do not work well for hiPSC culture because of cell agglomeration issues [29–31]. Other 3D cultures using scaffolds of natural polymers [32, 33] or hydrogel [34, 35] generated *in vivo*-like conditions for hiPSC culture to avoid cell agglomeration associated with the suspension system. Though the microcarrier bead approach in some of these studies has been considered the "3D" method for large-scale cell manufacturing using stirring tank bioreactors [29–31, 33], it is still based on 2D cell culture principles. In addition, natural polymer-derived scaffolds generally require a relatively complicated process for cell encapsulation and harvesting, resulting in significant cell loss that would hinder downstream analysis and applications. 3D bioprinting has been explored as a promising method for large-scale production of hiPSC spheroids which allows precise control of spheroid size, but photo-polymerization

[36–40] or temperature control [38, 41, 42] is often required for post-printing gelation. These conditions oftentimes can be harsh for human PSCs [43], leading to low viability and insufficient maintenance of pluripotency.

Hydrogel, on the other hand, is relatively easy to handle. Hydrogels consist of fully synthetic components that not only meet the need to create a well-defined culture environment but also allow the integration of bioactive molecules to promote cellular functions. Lei and Schaffer [35] developed polyethylene glycol (PEG) tailored hydrogel poly(*N*-isopropylacrylamide)-co-poly(ethylene glycol) (PNIPAAm-PEG or Mebiol Gel™), which is temperature-sensitive and starts gelation at 37°C. This hydrogel was reported to sustain a satisfactory growth rate for hESC and hiPSC lines [35] and maintained pluripotency. However, Mebiol Gel requires high seeding density (1×10^6 cells/ml), and is not easy to use due to its temperature-sensitive nature (it needs to be processed at very low or icy temperatures [35, 44, 45]). In addition, cell viability in Mebiol Gel was less than 20% [46], which is lower than required for various applications. Therefore, a practical and well-defined 3D matrix system that is proficient at maintaining high-quality hiPSCs for long-term culture and downstream applications remains to be developed.

2.2 Synthetic peptide hydrogel

A tri-block amphiphilic peptide-based hydrogel termed as h9e was discovered by Huang et al. [47] and Sun and Huang [48]. H9e is rationally designed from a group of selective amino acids that can self-assemble into nanofibers and then transform into a fast sol–gel reversible hydrogel through shearing force, such as pipetting or syringing, under neutral pH at room or body temperature [48, 49]. H9e's peptide sequence originated from two functional native proteins of human muscle [50] and the β-spiral motif of the spider flagelliform silk protein; therefore, it is highly compatible with biological systems [51]. This peptide is also reconcilable with various cell culture media, such as DMEM, MEM, RPMI, and L-15, as well as common hiPSC culture medium mTeSR and Essential 8 (E8). Through modification of the backbone structure, a variety of h9e can be created with the desirable hydrogel properties to meet specific requirements for 3D cell culture or *in vivo* delivery. For instance, PGmatrix system, a commercial product derived from h9e by PepGel LLC (Manhattan, KS), has been used for 3D cultures of various cancer cells [52–55] and hiPSC [46], *in vivo* delivery of drugs, antigens, viruses [47, 56, 57] as well as human MSCs [58] with PGmatrix were reported.

2.2.1 Mechanical properties of peptide hydrogel

PGmatrix is composed of entangled nanofibers forming a porous scaffold as shown in Atomic force microscopy (AFM) images (**Figure 1**). Single nanofiber with a 20 nm in diameter and nanofiber clusters with 100–500 nm in diameter were identified. The pore size was up to 2.5 μm. This peptide hydrogel is compatible with mTeSR1, the commonly used culture medium for hiPSCs, and formed a self-supporting hydrogel in a similar manner as reported by Liang et al. [55]. The self-assembling nature of PGmatrix was observed by measuring gel formation as a function of time. Within a few seconds, gel strength reached 100 Pa at the concentrations of 0.5% and 1.0% peptide; shear-thinning and self-recovery properties were measured at peptide concentrations of 0.2%, 0.5%, and 1.0% (**Figure 2**). After 1 minute of shear-thinning treatment, PGmatrix gel transformed into liquid-like status with about 0.1–0.5 Pa

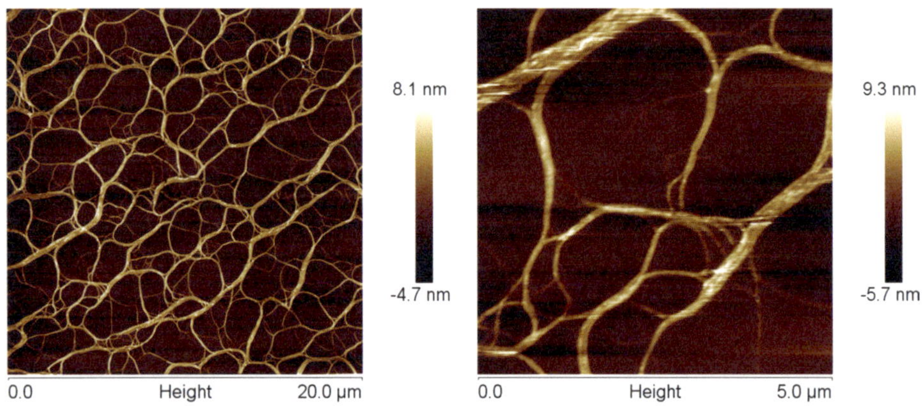

Figure 1.
AFM images of the PGmatrix nanostructure at low (left) and high (right) resolutions.

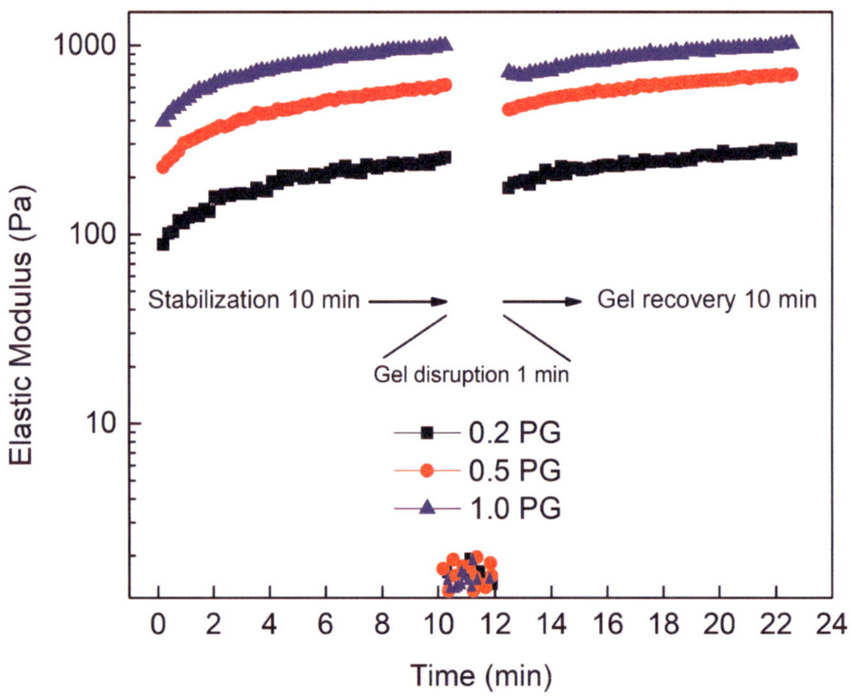

Figure 2.
Storage moduli of PGmatrix were directly proportional to the concentrations of peptide nanofibers in both shear-thinning and sol–gel recovery tests.

of elastic moduli. Once the shear-thinning force was removed, it quickly recovered to gel status. After 1 minute, the gel recovered to 67%, 70%, and 80% of its original strength in the samples with 0.2%, 0.5%, and 1.0% concentrations of peptide, respectively. After 10 minutes, all three samples restored up to 95% of the original gel strength (**Figure 2**). These unique properties suggest that PGmatrix scaffold can be easily manipulated manually or automatically by high throughput robot or bioprinting for cell encapsulation, expansion, and mechanical isolation.

2.2.2 Peptide hydrogel for long term hiPSC culture

hiPSCs derived from foreskin fibroblasts were cultured in 0.5% PGmatrix by Li et al. [46] in parallel with 2D culture in Matrigel-coated plates. hiPSCs grown in 3D PGmatrix showed significantly higher fold expansion ($p = 0.014$) and cell viability ($p < 0.0001$) (**Figure 3A**) compared to those on 2D Matrigel (**Figure 3B**). The variance of fold expansion across multiple passages (P13–P25) of hiPSC was significantly smaller in 3D (15.70 ± 3.70) than in 2D (10.55 ± 7.05) ($p < 0.05$). Another hiPSC line derived from CD44+ somatic cells was cultured in 3D PGmatrix as well as in 2D for multiple passages. Similarly, less variance in fold expansion and viability was observed with 3D culture [46].

After long-term culture in 3D PGmatrix, hiPSCs showed similar expression of Oct4, Nanog, Sox2, and SSEA4, which are comparable to those in 2D-cultured cells, however, differentiation marker gene AFP and Brachyury were significantly less expressed in 3D-cultured hiPSCs (**Figure 4**) [46]. The pluripotency of hiPSCs cultured in 3D PGmatrix was also verified with teratoma formation assay. In addition, these hiPSCs retained normal karyotype after long-term maintenance in 3D PGmatrix [46].

However, it is interesting to note that compared to the expression levels of SSEA4 and TRA-1-81 in pooled hPSC lines on 2D culture as reported by the International Stem Cell Initiative [59], TRA-1-81 expression (~7%) was significantly lower in two hiPSC lines cultured in 3D PGmatrix hydrogel. Although SSEA4 and TRA-1-81 have been commonly used as hiPSC markers, the TRA-1-81 expression may not be crucial in the maintenance of hiPSC pluripotency. It has been reported that naïve-state hPSCs did not express SSEA4 in both TRA-1-81 negative and positive fractions [60, 61]. Even though both SSEA4 and TRA-1-81 antigens are believed to be involved in cell adhesion, the mechanisms underlying hiPSC maintenance in 3D PGmatrix might be different from that of the conventional 2D culture. It is possible that high expression of SSEA4 and low expression of TRA-1-81 observed in the present 3D culture is due to direct physical or biological interactions between h9e peptide and hiPSCs. Nevertheless, the findings from this study suggest that some of the PSC markers established in 2D culture may need to be revised for the characterization of 3D-cultured hiPSCs.

On the other hand, differential expressions of certain genes were identified in 3D-cultured hiPSCs. Significant upregulation of UTF1 and hTERT, but

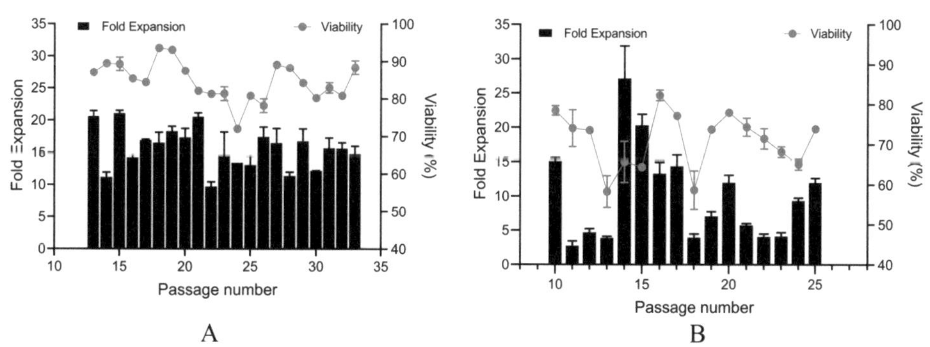

Figure 3.
Growth performance of hiPSCs in 3D PGmatrix (A) and in 2D on Matrigel (B). Data are shown as means ± SDs.

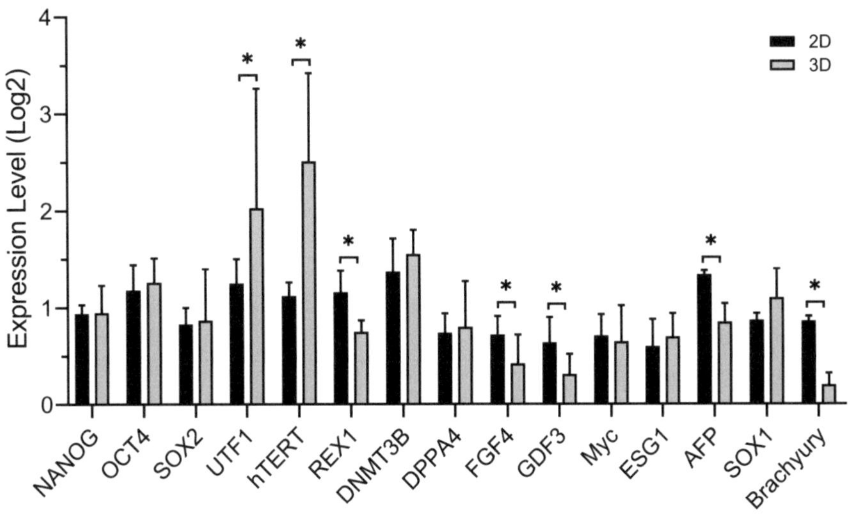

Figure 4.
*hiPSCs showed upregulation of pluripotency-related genes in 3D PGmatrix. 2D results were averaged from two different passages (P10 and P15), while 3D results were averaged from four different passages (P7, P10, P15, and P25). Data are shown as means ± SDs. *$p < 0.05$.*

downregulation of REX1, FGF4, and GDF3 were observed for both hiPSC lines in 3D compared to 2D culture (**Figure 4**). Being a PSC gene, UTF1 is closely associated with stem cell pluripotency. Researchers found that UTF1 may play an important role in chromatin formation of embryonic stem cell (ESC) [62]. Another critical factor controlling the proliferation of PSCs is the integrity of telomeres, which is regulated by hTERT transcriptor [63]. Overexpression of hTERT in hMSCs increased differentiation potential and decreased spontaneous differentiation [64]. Therefore, high expression levels of UTF1 and hTERT in the present 3D hiPSCs may imply high-quality hiPSCs in the aspects of genetic integrity, pluripotency, and proliferation compared to those in 2D systems. REX1 (also known as Zfp42) is a zinc-finger encoding gene expressed exclusively in early embryos and has been widely used as a PSC marker [65]. Early studies demonstrated that knockout of REX1 (REX1-/-) in ESCs does not affect cell proliferation and pluripotency [66, 67]. A later study discovered that REX1 regulates human stem cell pluripotency by promoting mitochondrial fission, which keeps mitochondria in an immature state and stem cells in a highly glycolytic state [68]. This function of REX1 is critical to protecting hiPSCs in 2D because a high oxygen level in 2D culture would trigger mitochondrial oxidative phosphorylation. However, in the 3D PGmatrix system, oxygen level might be limited by hydrogel as a "diffusion barrier"; therefore, upregulation of REX1 would be less critical for cells in 3D culture.

2.3 Comparison of hydrogels for 3D hiPSC culture

Since the rapid growth and high viability of hiPSCs cultured in 3D PGmatrix might be simply resulted from a 3D culture environment, 3D culture in PGmatrix was compared with PNIPAAm-PEG, Mebiol Gel [35] in parallel using the same hiPSCs using either mTeSR1 or E8 medium [46]. Under optimal conditions (PG-mTeSR1 vs. Mebiol-E8), hiPSCs formed a more uniform spherical morphology than those in Mebiol gel. Spheroids were larger in 3D Mebiol gel, but cells on the edge of the large spheroids seemed to be dying (**Figure 5**). Pluripotency gene expression compared

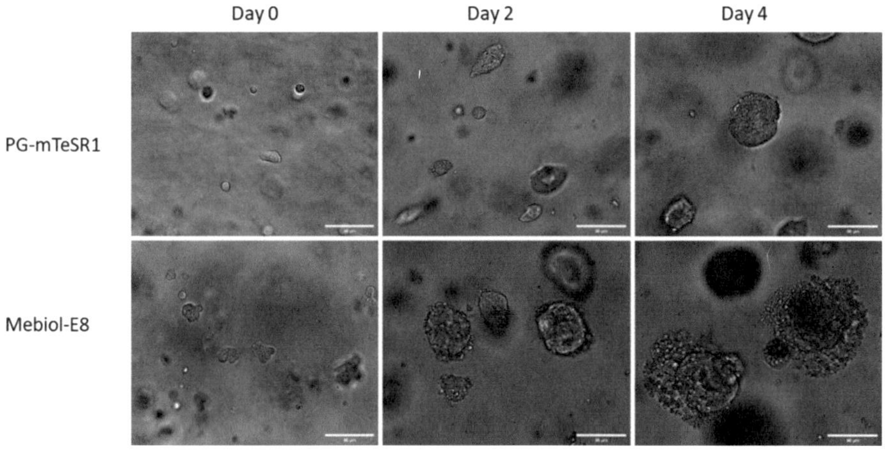

Figure 5.
Morphology of hiPSCs grown in PG-mTeSR1 and Mebiol-E8 on days 0, 2, and 4 after encapsulation. Scale bar, 50 µm.

by RT-qPCR showed that spheroids formed in the 3D Mebiol gel had lower pluripotency than those in the 3D PGmatrix [46]. These findings suggested that hiPSC maintenance performance is not only linked to 3D cell conformation and hypoxic condition provided by hydrogel. By comparing gel strength kinetics, gel degradability may be a contributing factor to hiPSC growth [46]. It has been proved that matrix degradation is crucial for the maintenance of neural progenitor cell stemness [69], the mechanosensing genes yes-associated protein 1 (YAP) and transcriptional coactivator with PDZ-binding motif (TAZ) [69] are at the center of this signaling cascade, which was also confirmed by Li et al. [46]. The maintenance of hiPSC pluripotency may be related to PGmatrix's degradability that allows encapsulated hiPSCs to modify their environment. Soluble factors such as insulin, bFGF, TGFβ, and nodal from medium supplements would bind to cell surface receptors such as GPCRs (G protein-coupled receptors) and RTKs (receptor tyrosine kinases) to activate essential pathways, including phosphoinositide-3-kinase–protein kinase B/Akt (PI3K/Akt) for survival [70, 71]. During this adaptation, hiPSCs may initiate their secretion of extracellular matrix (ECM) proteins and some proteases and start remodeling the surrounding environment through matrix degradation and modification. Secreted ECM proteins such as laminin and vitronectin would in turn bind to integrin and trigger various downstream signaling pathways to promote proliferation and pluripotency maintenance [72–74]. The microenvironment of ECM-protein-modified-PGmatrix would facilitate cell migration, leading to increased cell–cell contact *via* E-cadherin [75–77]. These binding events are closely linked to cell cytoskeleton, affecting actin dynamics and relaying the signals through Hippo pathway. Mechanical signals transduced through the signaling cascade, including LATS 1/2 would cause upregulation of the mechanosensitive YAP and TAZ proteins, which then relocate into the nucleus to affect gene expression related to pluripotency maintenance (**Figure 6**) [46, 78–82].

2.4 Develop peptide hydrogel for hiPSC bioprinting

To meet the need for 3D bioprinting, PGmatrix peptide hydrogel was modified into PGmatrix-M bioink that self-heals to a gel state after printing without any crosslinking

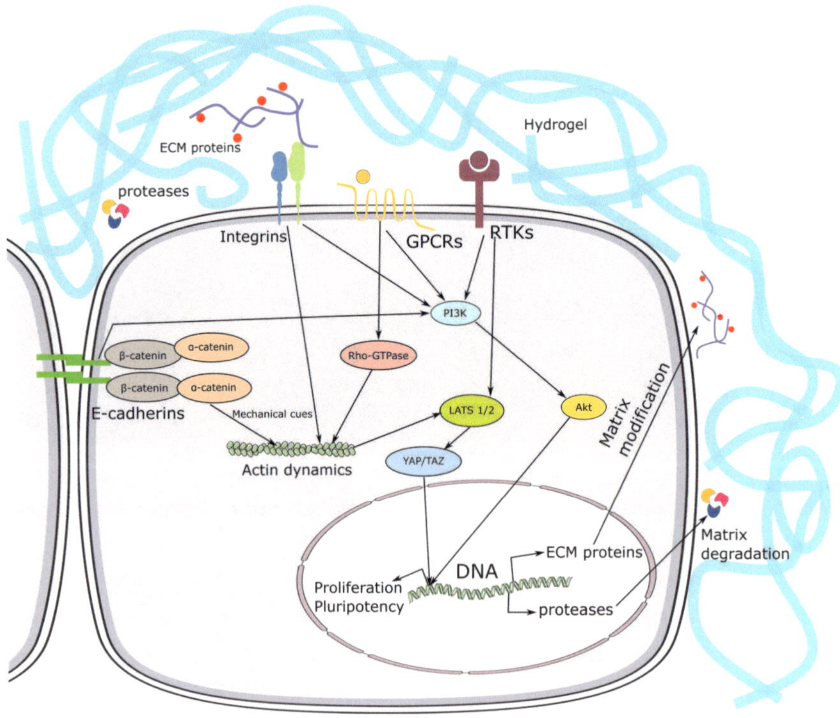

Figure 6.
Proposed mechanism for hiPSC growth and maintenance of pluripotency in 3D PGmatrix hydrogel. (Reprinted from Li et al. [46] with permission.)

aids from UV light or chemicals. In addition, PGmatrix-M supports hiPSC proliferation and maintenance after printing with viability above 95% [46]. It has been demonstrated that after 3D bioprinting, hiPSC aggregated can be differentiated into various tissues, including cartilage [83], hepatocyte-like cells [84], and neural tissues [85]. Compared to bioinks used in these studies, PGmatrix-M offered a gentler environment that allows long-term hiPSC maintenance. This raised the possibility of culture patterned hiPSC spheroids after printing and perform more complex differentiation processes, which would greatly benefit tissue engineering to produce functional organoids.

3. Conclusion

We have reviewed the advanced development in 3D hiPSC culture systems that have replaced the initial complex feeder layer or protein mixture Matrigel. Among the 3D technologies, the innovative peptide hydrogel PGmatrix outperformed its predecessors in terms of long-term hiPSC maintenance and the ability to be easily handled under ambient conditions. It is believed that usage of this 3D platform in the hiPSC field will not only promote the production of high-quality hiPSCs, which could lead to improvement of differentiation efficiency in generations of various types of somatic cells, including hepatocytes, but also provides a new tool to manufacture hiPSC at an industrial scale for downstream applications. In addition, the classical stem cell markers, such as TRA-1-81, should be reevaluated regarding their sensitivity in characterizing hiPSCs in 3D culture.

Conflict of interest

The authors declare no conflict of interest.

Appendices and nomenclature

AFP	alpha-fetoprotein
AFM	atomic force microscopy
bFGF	basic fibroblast growth factor
ECM	extracellular matrix
GPCRs	G protein-coupled receptors
hESCs	human embryonic stem cells
hiPSCs	human induced pluripotent stem cells
hPSCs	human pluripotent stem cells
MSCs	mesenchymal stem cells
MEF	mouse embryonic fibroblast
PI3K/Akt	phosphoinositide-3-kinase–protein kinase B/Akt
PEG	polyethylene glycol
PNIPAAm-PEG	poly(*N*-isopropylacrylamide)-co-poly(ethylene glycol)
RTKs	receptor tyrosine kinases
RT-qPCR	reverse transcription-quantitative polymerase chain reaction
ROCK	rho-associated, coiled-coil-containing protein kinase
3D	three-dimension
TGFβ	transforming growth factor-beta
TAZ	transcriptional coactivator with PDZ-binding motif
YAP	yes-associated protein 1

Author details

Quan Li, Guangyan Qi and Xiuzhi Susan Sun*
Department of Biological and Agricultural Engineering, Department of Grain Science and Industry, Kansas State University, Manhattan, Kansas, USA

*Address all correspondence to: xss@k-state.edu

References

[1] Zhu Z, Huangfu D. Human pluripotent stem cells: an emerging model in developmental biology. Development. 2013;**140**(4):705-717

[2] Takahashi K, Tanabe K, Ohnuki M, Narita M, Ichisaka T, Tomoda K, et al. Induction of pluripotent stem cells from adult human fibroblasts by defined factors. Cell. 2007;**131**(5):861-872

[3] Ohnuki M, Takahashi K. Present and future challenges of induced pluripotent stem cells. Philosophical Transactions of the Royal Society of London. Series B, Biological Sciences. 2015;**370**(1680):20140367

[4] Messina A, Luce E, Hussein M, Dubart-Kupperschmitt A. Pluripotent-stem-cell-derived hepatic cells: hepatocytes and organoids for liver therapy and regeneration. Cell. 2020;**9**(2):420

[5] Pettinato G, Lehoux S, Ramanathan R, Salem MM, He LX, Muse O, et al. Generation of fully functional hepatocyte-like organoids from human induced pluripotent stem cells mixed with Endothelial Cells. Scientific Reports. 2019;**9**(1):8920

[6] Tian L, Gao J, Garcia IM, Chen HJ, Castaldi A, Chen YW. Human pluripotent stem cell-derived lung organoids: potential applications in development and disease modeling. Wiley Interdisciplinary Reviews: Developmental Biology. 2021;**10**(6):e399

[7] Lee J, Rabbani CC, Gao H, Steinhart MR, Woodruff BM, Pflum ZE, et al. Hair-bearing human skin generated entirely from pluripotent stem cells. Nature. 2020;**582**(7812):399-404

[8] Brassard JA, Lutolf MP. Engineering stem cell self-organization to build better organoids. Cell Stem Cell. 2019;**24**(6):860-876

[9] Conner DA. Mouse embryo fibroblast (MEF) feeder cell preparation. Current Protocols in Molecular Biology. 2001;**51**(1)

[10] Kleinman HK, Martin GR. Matrigel: basement membrane matrix with biological activity. Seminars in Cancer Biology. 2005;**15**(5):378-386

[11] Miyazaki T, Futaki S, Hasegawa K, Kawasaki M, Sanzen N, Hayashi M, et al. Recombinant human laminin isoforms can support the undifferentiated growth of human embryonic stem cells. Biochemical and Biophysical Research Communications. 2008;**375**(1):27-32

[12] Miyazaki T, Futaki S, Suemori H, Taniguchi Y, Yamada M, Kawasaki M, et al. Laminin E8 fragments support efficient adhesion and expansion of dissociated human pluripotent stem cells. Nature Communications. 2012;**3**(1):1236

[13] Xu C, Inokuma MS, Denham J, Golds K, Kundu P, Gold JD, et al. Feeder-free growth of undifferentiated human embryonic stem cells. Nature Biotechnology. 2001;**19**(10):971-974

[14] Nagaoka M, Si-Tayeb K, Akaike T, Duncan SA. Culture of human pluripotent stem cells using completely defined conditions on a recombinant E-cadherin substratum. BMC Developmental Biology. 2010;**10**:60

[15] Rodin S, Antonsson L, Niaudet C, Simonson OE, Salmela E, Hansson EM, et al. Clonal culturing of human embryonic stem cells on

laminin-521/E-cadherin matrix in defined and xeno-free environment. Nature Communications. 2014;**5**:3195

[16] Lu J, Hou R, Booth CJ, Yang SH, Snyder M. Defined culture conditions of human embryonic stem cells. Proceedings of the National Academy of Sciences of the United States of America. 2006;**103**(15):5688-5693

[17] Braam SR, Zeinstra L, Litjens S, Ward-van Oostwaard D, van den Brink S, van Laake L, et al. Recombinant vitronectin is a functionally defined substrate that supports human embryonic stem cell self-renewal via alphavbeta5 integrin. Stem cells (Dayton, Ohio). 2008;**26**(9):2257-2265

[18] Lambshead JW, Meagher L, Goodwin J, Labonne T, Ng E, Elefanty A, et al. long-term maintenance of human pluripotent stem cells on cRGDfK-presenting synthetic surfaces. Scientific Reports. 2018;**8**:701

[19] Villa-Diaz LG, Nandivada H, Ding J, Nogueira-de-Souza NC, Krebsbach PH, O'Shea KS, et al. Synthetic polymer coatings for long-term growth of human embryonic stem cells. Nature Biotechnology. 2010;**28**(6):581-583

[20] Irwin EF, Gupta R, Dashti DC, Healy KE. Engineered polymer-media interfaces for the long-term self-renewal of human embryonic stem cells. Biomaterials. 2011;**32**(29):6912-6919

[21] Brafman DA, Chang CW, Fernandez A, Willert K, Varghese S, Chien S. Long-term human pluripotent stem cell self-renewal on synthetic polymer surfaces. Biomaterials. 2010;**31**(34):9135-9144

[22] Dang LTH, Feric NT, Laschinger C, Chang WY, Zhang B, Wood GA, et al. Inhibition of apoptosis in human induced pluripotent stem cells during expansion in a defined culture using angiopoietin-1 derived peptide QHREDGS. Biomaterials. 2014;**35**(27):7786-7799

[23] Derda R, Li L, Orner BP, Lewis RL, Thomson JA, Kiessling LL. Defined substrates for human embryonic stem cell growth identified from surface arrays. ACS Chemical Biology. 2007;**2**(5):347-355

[24] Klim JR, Li L, Wrighton PJ, Piekarczyk MS, Kiessling LL. A defined glycosaminoglycan-binding substratum for human pluripotent stem cells. Nature Methods. 2010;**7**(12):989-994

[25] Melkoumian Z, Weber JL, Weber DM, Fadeev AG, Zhou Y, Dolley-Sonneville P, et al. Synthetic peptide-acrylate surfaces for long-term self-renewal and cardiomyocyte differentiation of human embryonic stem cells. Nature Biotechnology. 2010;**28**(6):606-610

[26] Centeno EGZ, Cimarosti H, Bithell A. 2D versus 3D human induced pluripotent stem cell-derived cultures for neurodegenerative disease modelling. Molecular Neurodegeneration. 2018;**13**(1):27

[27] Zhou Y, Chen H, Li H, Wu Y. 3D culture increases pluripotent gene expression in mesenchymal stem cells through relaxation of cytoskeleton tension. Journal of Cellular and Molecular Medicine. 2017;**21**(6):1073

[28] Bijonowski BM, Fu Q, Yuan X, Irianto J, Li Y, Grant SC, et al. Aggregation-induced integrated stress response rejuvenates culture-expanded human mesenchymal stem cells. Biotechnology and Bioengineering. 2020;**117**(10):3136-3149

[29] Chen AKL, Chen X, Choo ABH, Reuveny S, Oh SKW. Critical microcarrier properties affecting the expansion of undifferentiated human embryonic stem cells. Stem Cell Research. 2011;7(2):97-111

[30] Chen VC, Couture SM, Ye J, Lin Z, Hua G, Huang HIP, et al. Scalable GMP compliant suspension culture system for human ES cells. Stem Cell Research. 2012;**8**(3):388-402

[31] Serra M, Correia C, Malpique R, Brito C, Jensen J, Bjorquist P, et al. Microencapsulation technology: A powerful tool for integrating expansion and cryopreservation of human embryonic stem cells. PLoS One. 2011;**6**(8):e23212

[32] Li Z, Leung M, Hopper R, Ellenbogen R, Zhang M. Feeder-free self-renewal of human embryonic stem cells in 3D porous natural polymer scaffolds. Biomaterials. 2010;**31**(3):404-412

[33] Chayosumrit M, Tuch B, Sidhu K. Alginate microcapsule for propagation and directed differentiation of hESCs to definitive endoderm. Biomaterials. 2010;**31**(3):505-514

[34] Gerecht S, Burdick JA, Ferreira LS, Townsend SA, Langer R, Vunjak-Novakovic G. Hyaluronic acid hydrogel for controlled self-renewal and differentiation of human embryonic stem cells. Proceedings of the National Academy of Sciences of the United States of America. 2007;**104**(27):11298-11303

[35] Lei Y, Schaffer DV. A fully defined and scalable 3D culture system for human pluripotent stem cell expansion and differentiation. Proceedings of the National Academy of Sciences of the United States of America. 2013;**110**(52):E5048

[36] Gao G, Yonezawa T, Hubbell K, Dai G, Cui X. Inkjet-bioprinted acrylated peptides and PEG hydrogel with human mesenchymal stem cells promote robust bone and cartilage formation with minimal printhead clogging. Biotechnology Journal. 2015;**10**(10):1568-1577

[37] Poldervaart MT, Goversen B, de Ruijter M, Abbadessa A, Melchels FPW, Öner FC, et al. 3D bioprinting of methacrylated hyaluronic acid (MeHA) hydrogel with intrinsic osteogenicity. PLoS One. 2017;**12**(6):e0177628

[38] Suntornnond R, Tan EYS, An J, Chua CK. A highly printable and biocompatible hydrogel composite for direct printing of soft and perfusable vasculature-like structures. Scientific Reports. 2017;**7**(1):16902

[39] Shi L, Carstensen H, Hölzl K, Lunzer M, Li H, Hilborn J, et al. Dynamic Coordination Chemistry Enables Free Directional Printing of Biopolymer Hydrogel. Chemistry of Materials. 2017;**29**(14):5816-5823

[40] Skardal A, Devarasetty M, Kang HW, Mead I, Bishop C, Shupe T, et al. A hydrogel bioink toolkit for mimicking native tissue biochemical and mechanical properties in bioprinted tissue constructs. Acta Biomaterialia. 2015;**25**:24-34

[41] Giuseppe MD, Law N, Webb B, Macrae RA, Liew LJ, Sercombe TB, et al. Mechanical behaviour of alginate-gelatin hydrogels for 3D bioprinting. Journal of the Mechanical Behavior of Biomedical Materials. Mar 2018;**79**:150-157

[42] Hull SM, Lindsay CD, Brunel LG, Shiwarski DJ, Tashman JW, Roth JG, et al. 3D Bioprinting using UNIversal Orthogonal Network (UNION) bioinks. Advanced Functional Materials. 2021;**31**(7):2007983

[43] Salaris F, Rosa A. Construction of 3D in vitro models by bioprinting human pluripotent stem cells: challenges and opportunities. Brain Research. 2019;**1723**:146393

[44] Adil MM, Rodrigues GMC, Kulkarni RU, Rao AT, Chernavsky NE, Miller EW, et al. Efficient generation of hPSC-derived midbrain dopaminergic neurons in a fully defined, scalable, 3D biomaterial platform. Scientific Reports. 2017;**7**:40573

[45] Adil MM, Vazin T, Ananthanarayanan B, Rodrigues GMC, Rao AT, Kulkarni RU, et al. Engineered hydrogels increase the post-transplantation survival of encapsulated hESC-derived midbrain dopaminergic neurons. Biomaterials. 2017;**136**:1-11

[46] Li Q, Qi G, Liu X, Bai J, Zhao J, Tang G, et al. Universal peptide hydrogel for scalable physiological formation and bioprinting of 3D spheroids from human induced pluripotent stem cells. Advanced Functional Materials. 2021;**31**(41):2104046

[47] Huang H, Shi J, Laskin J, Liu Z, McVey DS, Sun XS. Design of a shear-thinning recoverable peptide hydrogel from native sequences and application for influenza H1N1 vaccine adjuvant. Soft Matter. 2011;**7**(19):8905-8912

[48] Sun XS, Huang H. Protein peptide hydrogels [Internet]. US8835395B2, 2014 [cited 2022]. Available from: https://patents.google.com/patent/US8835395B2/en

[49] Huang H, Herrera AI, Luo Z, Prakash O, Sun XS. Structural transformation and physical properties of a hydrogel-forming peptide studied by NMR, transmission electron microscopy, and dynamic rheometer. Biophysical Journal. 2012;**103**(5):979-988

[50] Shen X, Mo X, Moore R, Frazier SJ, Iwamoto T, Tomich JM, et al. Adhesion and structure properties of protein nanomaterials containing hydrophobic and charged amino acids. Journal of Nanoscience and Nanotechnology. 2006;**6**(3):837-844

[51] Huang H, Sun XS. Rational design of responsive self-assembling peptides from native protein sequences. Biomacromolecules. 2010;**11**(12):3390-3394

[52] Huang H, Ding Y, Sun XS, Nguyen TA. Peptide hydrogelation and cell encapsulation for 3D culture of MCF-7 breast cancer cells. PLoS One. 2013;**8**(3):e59482

[53] Kumar D, Kandl C, Hamilton CD, Shnayder Y, Tsue TT, Kakarala K, et al. Mitigation of tumor-associated fibroblast-facilitated head and neck cancer progression with anti–hepatocyte growth factor antibody Ficlatuzumab. JAMA Otolaryngol-Head & Neck Surgery. 2015;**141**(12):1133-1139

[54] Miller PG, Shuler ML. Design and demonstration of a pumpless 14 compartment microphysiological system. Biotechnology and Bioengineering. 2016;**113**(10):2213-2227

[55] Liang J, Susan Sun X, Yang Z, Cao S. Anticancer drug Camptothecin test in 3d hydrogel networks with HeLa cells. Scientific Reports. 2017;**7**:37626

[56] Li X, Galliher-Beckley A, Huang H, Sun X, Shi J. Peptide nanofiber hydrogel adjuvanted live virus vaccine enhances cross-protective immunity to porcine reproductive and respiratory syndrome virus. Vaccine. 2013;**31**(41):4508-4515

[57] Liang J, Liu G, Wang J, Sun XS. Controlled release of BSA-linked cisplatin through a PepGel self-assembling peptide

nanofiber hydrogel scaffold. Amino Acids. 2017;**49**(12):2015-2021

[58] Li Q, Qi G, Lutter D, Beard W, Souza CRS, Highland MA, et al. Injectable peptide hydrogel encapsulation of mesenchymal stem cells improved viability, stemness, anti-inflammatory effects, and early stage wound healing. Biomolecules. 2022;**12**(9):1317

[59] International Stem Cell Initiative, Adewumi O, AflatoonianB, Ahrlund-RichterL, AmitM, Andrews PW, et al. Characterization of human embryonic stem cell lines by the International Stem Cell Initiative. Nature Biotechnology. 2007;**25**(7):803-816

[60] Brimble SN, Sherrer ES, Uhl EW, Wang E, Kelly S, Merrill AH, et al. The cell surface glycosphingolipids SSEA-3 and SSEA-4 are not essential for human ESC pluripotency. Stem cells (Dayton, Ohio). 2007;**25**(1):54-62

[61] Pastor WA, Chen D, Liu W, Kim R, Sahakyan A, Lukianchikov A, et al. Naive human pluripotent cells feature a methylation landscape devoid of blastocyst or germline memory. Cell Stem Cell. 2016;**18**(3):323-329

[62] Morshedi A, Soroush Noghabi M, Dröge P. Use of UTF1 genetic control elements as iPSC reporter. Stem Cell Reviews and Reports. 2013;**9**(4):523-530

[63] Ramlee MK, Wang J, Toh WX, Li S. Transcription regulation of the human telomerase reverse transcriptase (hTERT) gene. Genes. 2016;**7**(8):50

[64] Tsai CC, Chen CL, Liu HC, Lee YT, Wang HW, Hou LT, et al. Overexpression of hTERT increases stem-like properties and decreases spontaneous differentiation in human mesenchymal stem cell lines. Journal of Biomedical Science. 2010;**17**(1):64

[65] Lee KC, Wong WK, Feng B. Decoding the pluripotency network: the emergence of new transcription factors. Biomedicine. 2013;**1**(1):49-78

[66] Masui S, Ohtsuka S, Yagi R, Takahashi K, Ko MSH, Niwa H. Rex1/Zfp42 is dispensable for pluripotency in mouse ES cells. BMC Developmental Biology. 2008;**8**:45

[67] Scotland KB, Chen S, Sylvester R, Gudas LJ. Analysis of Rex1 (zfp42) function in embryonic stem cell differentiation. Developmental Dynamics. 2009;**238**(8):1863-1877

[68] Son MY, Choi H, Han YM, Sook CY. Unveiling the critical role of REX1 in the regulation of human stem cell pluripotency. Stem Cells. 2013;**31**(11):2374-2387

[69] Madl CM, LeSavage BL, Dewi RE, Dinh CB, Stowers RS, Khariton M, et al. Maintenance of neural progenitor cell stemness in 3D hydrogels requires matrix remodeling. Nature Materials. 2017;**16**(12):1233-1242

[70] Hemmings BA, Restuccia DF. PI3K-PKB/Akt pathway. Cold Spring Harbor Perspectives in Biology. 2012;**4**(9):a011189

[71] Hossini AM, Quast AS, Plötz M, Grauel K, Exner T, Küchler J, et al. PI3K/AKT signaling pathway is essential for survival of induced pluripotent stem cells. PLoS One. 2016;**11**(5):e0154770

[72] Laperle A, Hsiao C, Lampe M, Mortier J, Saha K, Palecek SP, et al. α-5 Laminin synthesized by human pluripotent stem cells promotes self-renewal. Stem Cell Reports. 2015;**5**(2):195-206

[73] Xu Y, Zhu X, Hahm HS, Wei W, Hao E, Hayek A, et al. Revealing a core signaling regulatory mechanism for

pluripotent stem cell survival and self-renewal by small molecules. Proceedings of the National Academy of Sciences of the United States of America. 2010;**107**(18):8129-8134

[74] Hayashi Y, Furue MK. Biological effects of culture substrates on human pluripotent stem cells. Stem Cells International. 2016;**2016**:5380560

[75] Soncin F, Ward CM. The function of e-cadherin in stem cell pluripotency and self-renewal. Genes. 2011;**2**(1):229-259

[76] De Santis G, Miotti S, Mazzi M, Canevari S, Tomassetti A. E-cadherin directly contributes to PI3K/AKT activation by engaging the PI3K-p85 regulatory subunit to adherens junctions of ovarian carcinoma cells. Oncogene. 2009;**28**(9):1206-1217

[77] Li L, Bennett SAL, Wang L. Role of E-cadherin and other cell adhesion molecules in survival and differentiation of human pluripotent stem cells. Cell Adhesion & Migration. 2012;**6**(1):59-70

[78] Shao Y, Sang J, Fu J. On human pluripotent stem cell control: the rise of 3D bioengineering and mechanobiology. Biomaterials. 2015;**52**:26-43

[79] McKee C, Brown C, Chaudhry GR. Self-Assembling scaffolds supported long-term growth of human primed embryonic stem cells and upregulated core and naïve pluripotent markers. Cell. 2019;**8**(12):1650

[80] Brusatin G, Panciera T, Gandin A, Citron A, Piccolo S. Biomaterials and engineered microenvironments to control YAP/TAZ-dependent cell behaviour. Nature Materials. 2018;**17**(12):1063-1075

[81] Moya IM, Halder G. Hippo-YAP/TAZ signalling in organ regeneration and regenerative medicine. Nature Reviews. Molecular Cell Biology. 2019;**20**(4):211-226

[82] Lee S, Stanton AE, Tong X, Yang F. Hydrogels with enhanced protein conjugation efficiency reveal stiffness-induced YAP localization in stem cells depends on biochemical cues. Biomaterials. 2019;**202**:26-34

[83] Nguyen D, Hägg DA, Forsman A, Ekholm J, Nimkingratana P, Brantsing C, et al. Cartilage tissue engineering by the 3D bioprinting of iPS cells in a nanocellulose/alginate bioink. Scientific Reports. 2017;**7**:658

[84] Faulkner-Jones A, Fyfe C, Cornelissen DJ, Gardner J, King J, Courtney A, et al. Bioprinting of human pluripotent stem cells and their directed differentiation into hepatocyte-like cells for the generation of mini-livers in 3D. Biofabrication. 2015;**7**(4):044102

[85] Abelseth E, Abelseth L, De la Vega L, Beyer ST, Wadsworth SJ, Willerth SM. 3D printing of neural tissues derived from human induced pluripotent stem cells using a fibrin-based bioink. ACS Biomaterials Science & Engineering. 2019;**5**(1):234-243

Section 4

Human Pluripotent Stem Cells for Mechanism Research

Chapter 7

The Potential of Human Induced Pluripotent Stem Cells (hiPSCs) for the Study of Channelopathies: Advances and Future Directions

Paul Disse, Nadine Ritter, Nathalie Strutz-Seebohm and Guiscard Seebohm

Abstract

Human induced pluripotent stem cells (hiPSCs) have revolutionized research on ion channels and channelopathies. Channelopathies are a group of genetic disorders characterized by dysfunctional ion channels, which are responsible for the regulation of ion flow across cell membranes. These disorders can affect various organ systems, leading to a wide range of symptoms and clinical manifestations. Differentiating pluripotent stem cells into various cell types results in the possibility of creating tissue- and disease-specific cell models. These models offer the possibility to investigate the underlying mechanisms of channelopathies and develop potential therapies. Using hiPSC-derived cells has allowed crucial insights into diseases like epilepsy, long QT syndrome, and periodic paralysis. However, the full potential of hiPSCs in this field is still to be exploited. The research will most likely focus on developing more complex cell models to further investigate channel dysfunction and its pathological consequences. In addition, hiPSCs will be increasingly used in drug screening and developing personalized therapies for various diseases. This chapter outlines the past and present achievements of hiPSCs in the field of channelopathies as well as provides an outlook on future possibilities.

Keywords: neurology, cardiology, channelopathies, epilepsy, long QT syndrome, medicine, diseases, treatment

1. Introduction

1.1 Definition and classification of channelopathies

Channelopathies are a group of genetic disorders characterized by dysfunctional ion channels, which play a crucial role in the regulation of ion flow across cell membranes. These disorders can affect various organ systems, leading to a wide range of symptoms and clinical manifestations. Understanding the definition and classification of channelopathies is essential for accurate diagnosis, appropriate management, and the development of targeted therapies [1].

Channelopathies can be classified based on several criteria, including the affected organ system, the type of ion channel involved, and the specific clinical manifestations, for example, neuronal, cardiac, and musculoskeletal channelopathies.

1.1.1 Neurological channelopathies

These channelopathies affect the nervous system, resulting in various neurological symptoms and disorders. For instance, voltage-gated sodium and potassium channel mutations can cause epilepsy, characterized by recurrent seizures and other dysfunctions of the cognitive, emotional, and neurological systems [2–4]. These seizures occur due to excessive and synchronous activity of nerve cells in the brain. Epilepsy can have various causes, including genetic predisposition, brain injury, infection, and metabolic disorders. Other neurological channelopathies include episodic ataxias, which manifest as episodes of uncoordinated movement and balance problems, and periodic paralysis, characterized by episodes of muscle weakness or paralysis [5–7].

1.1.2 Cardiac channelopathies

These channelopathies primarily affect the electrical properties of the heart, leading to arrhythmias and sudden cardiac death. Examples include Brugada syndrome, catecholaminergic polymorphic ventricular tachycardia (CPVT), familial atrial fibrillation, and long QT syndrome (LQTS) [8, 9]. The latter, long QT syndrome (LQTS), is a rare congenital disorder of the heart characterized by a delayed repolarization of the ventricles. It is characterized by prolongation of the QT interval time in the electrocardiogram (ECG). The QT interval represents the time taken to complete the depolarization and repolarization of the ventricles of the heart.

Long QT syndrome can be caused by genetic mutations that affect the function of the ion channels responsible for repolarization of the heart. However, medications such as antipsychotics can also lead to long QT syndrome. Thus, LQTS can be inherited and/or acquired [8, 10].

One of the most serious complications of long QT syndrome is torsade de pointes tachycardia, a specific form of ventricular tachycardia in which the heart beats irregularly and rapidly. It is characterized by a twisting of the QRS complex around the isoelectric line in the ECG. Torsade de pointes is a life-threatening cardiac arrhythmia, that can transform into ventricular fibrillation, which may lead to sudden cardiac death [10–12].

1.1.3 Musculoskeletal channelopathies

One example of a skeletal muscle channelopathy is myotonia congenita, which is characterized by muscle stiffness and delayed muscle relaxation after contraction. This condition is caused by mutations in chloride or sodium channels that affect the electrical properties of skeletal muscles [13, 14]. Similarly, periodic paralysis, as mentioned earlier, can also affect the muscles and lead to episodes of muscle weakness or paralysis [6, 7].

Neurological, cardiac, and musculoskeletal channelopathies are some of the commonly classified types, but many organ systems can be affected by channelopathies, also at the same time. The clinical manifestations can vary greatly depending on the mutation and its effects on ion channels. Cooperation between researchers, neurologists, cardiologists, and other specialists is crucial for the diagnosis, management, and treatment of channelopathies.

1.1.4 Genetic mutations

Channelopathies arise due to the dysregulation of ion channels, which can be caused by various mechanisms.

The primary cause of channelopathies are genetic mutations that can affect the structure and function of ion channel proteins. These mutations can result in the proteins' loss-of-function or gain-of-function, leading to alterations in ion channel activity. For instance, missense mutations can cause changes in the amino acid sequence of ion channels, leading to altered channel gating, conductance, or trafficking. These genetic mutations can be inherited in an autosomal dominant or recessive manner, or they may arise *de novo* [15–17].

Genetic mutations can also lead to alterations in the channels' biophysical properties. Mutations can for example affect channel gating, resulting in changes in the voltage or ligand dependence of channel gating. Mutations can also affect the conductance or ion selectivity of a channel, leading to altered ion flux. These biophysical alterations can disrupt the normal electrical signaling and ion homeostasis in cells and tissues, contributing to the pathogenesis of channelopathies [18–20].

Further, the subcellular localization of a channel can be altered by a mutation causing dysregulations in the cell biology of the respective cell. Altered cellular and/or developmental functions can cause, for example, syndactyly and bone malformation due to impaired apoptosis in Andersen-Tawil syndrome or Timothy syndrome [21, 22].

Loss-of-function mutations of ion channels can lead to reduced or absent ion channel activity, impairing the regular flow of ions across cell membranes. This can disrupt cellular excitability, neurotransmission, and other essential physiological processes [18].

Gain-of-function mutations of ion channels can lead to increased ion channel activity or altered ion selectivity. This can result in excessive ion flux, aberrant electrical signaling, and cellular dysfunction, contributing to the development of channelopathies [23].

In some channelopathies, genetic mutations can also impair the proper trafficking and localization of ion channels to the cell membrane. Mutations can disrupt the interaction of channels with chaperones, affecting their folding, assembly, and transport to the plasma membrane. As a result, the number of functional channels at the cell surface may be reduced, leading to decreased ion conductance and altered cellular excitability [24, 25].

1.1.5 Altered channel regulation by modulators

Ion channel activity can be modulated by various endogenous or exogenous factors, including ligands, second messengers, and posttranslational modifications. In channelopathies, mutations can disrupt the normal regulation of ion channels by these modulators. For example, mutations can alter the sensitivity of channels to voltage changes, impair the binding of ligands or regulatory proteins, or disrupt the phosphorylation or glycosylation sites critical for channel function. These alterations in channel regulation can lead to abnormal ion channel activity and subsequent disease manifestations [26, 27].

1.1.6 Interactions with auxiliary subunits or interacting proteins

Ion channels interact with auxiliary subunits or regulatory proteins that modulate their activity or regulate their cellular localization. Mutations in these auxiliary subunits or interacting proteins can disrupt the normal function of ion channels. For example, mutations in β subunits of voltage-gated calcium channels can affect channel trafficking, gating, or modulation by intracellular signaling molecules. Similarly, mutations in accessory proteins involved in potassium channel function, such as KCNE subunits, can impair channel activity and lead to channelopathies [28, 29].

1.1.7 Multifactorial interactions

Channelopathies can also arise from complex interactions between genetic factors, environmental triggers, and other modifying factors. These interactions can influence the severity, onset, and progression of channelopathies [1, 6].

Finally, the promotor regions of ion channels couple ion channel genesis to promotor regulations and mutations in the promotors or in the transcription factors can disrupt controlled ion channel protein generation to cause complex disease states.

1.2 Definition and properties of stem cells

Stem cells are a unique type of cells that have the potential to regenerate themselves and differentiate into different cell types in the body [30]. They play a significant role during the development of the organism, as well as in the regeneration and repair of tissues and organs in the adult body [31]. The ability to self-renew and differentiate makes stem cells an important field of research with great potential for medical research and treatment.

Stem cells are divided into two main categories: embryonic stem cells (ES cells) and adult stem cells. ES cells originate from the inner cell mass of an embryo at a very early stage of development. They are pluripotent, which means that they have the potential to differentiate into virtually all cell types of the body [30]. Adult stem cells, on the other hand, are present in the adult body and can be found in specific tissues and organs. They are multipotent and have the potential to differentiate into various cell types within their tissue [32].

ES cells are often derived from supernumerary embryos created through assisted reproductive techniques. However, the derivation of ES cells results in the destruction of the embryo. Therefore, alternative sources of pluripotent stem cells have been explored, such as induced pluripotent stem cells. iPS cells are generated by reprogramming adult somatic cells, such as skin cells, into a pluripotent state [33]. This technique, developed by Shinya Yamanaka and his team, allows pluripotent stem cells to be derived without the need for embryo destruction [34] and has earned the Nobel Prize for Physiology and Medicine in 2012.

The properties of stem cells make them a valuable tool for regenerative medicine, disease research, and the development of new therapies. Due to their ability to self-renew, stem cells can be grown in large amounts to provide sufficient cell quantities for therapeutic applications. Their ability to differentiate enables the production of specific cell types needed for transplantation or tissue regeneration [31].

The field of stem cell research has revolutonized disease modeling. Stem cells are used to treat diseases such as blood disorders, degenerative diseases of the nervous system, and heart disease [35, 36]. The development of iPS cells has also opened up new possibilities for personalized medicine and the study of disease mechanisms [37–39].

2. Stem cell-based models of channelopathies

As described above, channelopathies lead to serious diseases or syndromes in patients. Studying these disorders and developing effective treatments can be challenging due to the limited availability of patient samples and the complex nature of affected tissues. However, recent advancements in stem cell research have opened new avenues for modeling channelopathies in the laboratory under controlled settings. Stem cell-based models provide valuable tools to investigate the underlying mechanisms and pathophysiology, as well as potential therapeutic strategies for channelopathies.

2.1 iPS-derived cells for physiological channel research and pathophysiological research

For the characterization of basic channel activity, generic cells such as Chinese Hamster ovary (CHO) or Human embryonic kidney (HEK) cells are broadly utilized [23]. These cells can be manipulated by transfection with cDNA constructs of channels or receptors. After overexpression of these proteins, the ion channel function in the cells can be examined using the patch clamp technique to record channel activity [40].

An alternative method for examining channel activity is Two-Electrode Voltage Clamp (TEVC). This involves injecting mRNA encoding the channels of interest into the cytosol of, that is, *Xenopus laevis* frog eggs, which are then expressed and recorded [41]. In both the patch clamp technique and TEVC, the channels can be examined by applying voltages or currents. In addition, it is possible to measure the effect of activators and inhibitors on the channels and to compare wild-type and mutant channels [23].

Despite the progress made by these established methods, there are reasons to turn to new models. The main reason is the physiological relevance of the cells used. While the cell models mentioned above investigate neuronal, cardiac, and other channels, the cells used are tumor cells or oocyte cells. Moreover, two of the cell systems mentioned are not even of human origin. However, primary cells from humans are often not available for regular lab work. The advantage of using induced pluripotent stem cells (iPS) is that they can be generated from any tissue sample of the human body [33]. Thus, truly neuronal channels can be studied in neuronal cells and cardiac channels in cardiac cells. This allows the channels to be studied in a much more physiological context compared to the previously used cell systems [42].

The use of patient-derived or transgenic cell lines also allows the study of pathophysiological mechanisms of channelopathies [43]. By studying channels in cells from patients with known channelopathies, the effects of disease mutations on channel function can be investigated to gain a better understanding of the pathophysiology of these diseases.

2.2 Disease modeling and drug screening

In present and future stem cell research, a variety of established differentiation protocols are available that enable precise differentiation of stem cells in both 2D and 3D cultures [44–46]. These protocols have proven to reliably generate neuronal and cardiac cell tissues and enable the targeted generation of various electrophysiological phenotypes.

For the differentiation of neuronal cell tissues, optimized protocols have been developed that allow stem cells to differentiate into specific neuronal cell types such as neurons, astrocytes, and oligodendrocytes. By combining specific growth factors and culture conditions, the desired electrophysiological properties can be studied [46–48].

Remarkable progress has also been made in the field of cardiac differentiation. The application of targeted protocols enables the efficient differentiation of stem cells into cardiac cell types such as working myocardium, atrial cells, and pacemakers. The specific combination of signals and factors can achieve the desired electrophysiological expression, which is important for the study of cardiac physiology and channelopathies [49–51].

These differentiated neuronal and cardiac cell tissues provide the opportunity to explore physiological channel activity in more detail and to study specific electrophysiological phenotypes. By using these stem cell-derived cell systems, the understanding of underlying mechanisms of channels in physiological processes is broadened. Using common and modern molecular biology methods, most notably CRISPR/Cas9, the cell lines can be modified to express channel pathologies [52]. Thus, new therapeutic approaches can be developed.

Stem cell-derived models of diseases like channelopathies offer a platform to investigate pathophysiological mechanisms at the cellular and molecular levels. By differentiating patient-specific iPSCs into disease-relevant cell types, researchers can analyze the functional consequences of the specific mutations on ion channel activity, cellular excitability, and downstream signaling pathways [53].

In the field of drug screening and safety pharmacology, stem cell-derived tissues are also playing an increasingly important role. In conventional screening, often transfected cancer cell systems or animal tissues are used, such as in patch clamps in HEK cells or mouse brain slices [54]. Stem cells also offer the possibility of drug screening in various derived cell identities to identify and explore potential candidate compounds that could alleviate disease symptoms or counteract disease progression.

However, in transfected cell systems, often only the effects on a specific channel type can be tested. In contrast, more physiologically complex cells and tissues, as derived from stem cells, offer the possibility to study the effects of drugs on an entire cell system.

In contrast to the well-established mouse brain slices, the hiPSCs offer another advantage: They are of human origin. And therefore closer to human (patho) physiology.

These advances in differentiating stem cells into specific cell types result in new possibilities for drug screening and safety pharmacology. By using stem cell-derived tissues, more comprehensive studies can be conducted on the effects of drugs on complex cell systems, leading to improved prediction of effects on the human body. This is particularly relevant for testing the efficacy and safety of potential therapeutic compounds in a disease-relevant context [48, 55].

Since all human tissues are formed from the same stem cell with the help of different differentiation protocols, the specific effects on different tissues can also be investigated, especially in the case of channel mutations. In a broader sense, the use of drugs and channel modulators on different tissue types can also be studied simultaneously.

2.3 Generation of patient-specific stem cells and personalized medicine

A significant advance in patient-specific research has been the development of techniques to generate stem cells derived from patients' tissues. Patient-derived iPS cells are generated by reprogramming adult somatic cells from a diseased patient. This reprogramming enables the phenotypically and genotypically pathological cells to be differentiated into a wide variety of cell types [34].

The use of patient-derived iPS cells offers several advantages. Firstly, it allows the generation of stem cells that are genetically matched to the patient. This means that patient-derived iPS cells carry the patient's genetic characteristics and disease mutations, which opens up the possibility of examining the properties of the patients' cells in iPS-derived tissues of any kind [56]. As shown in the recent past, these tissues have successfully been used for modeling and studying various diseases, including neurological diseases such as Parkinson's, Alzheimer's, and Amyotrophic Lateral Sclerosis (ALS) [46, 56–58].

Cells generated from patient material are the closest pathophysiological cell state, with a 100% genotype of a patient. These patient-derived stem cells have the potential to revolutionize personalized medicine and the study of diseases.

Secondly, these cells can be used to create patient-specific cell models that can be used to study disease processes and develop tailored therapies [59].

In addition, patient-derived iPS cells can be used for personalized medicine. They enable the development of patient-specific therapies in which drugs can be tested for efficacy and safety in the patient's individual cells [60]. This approach aims to improve the efficiency of medicines and reduce potentially harmful side effects [61, 62].

Especially for rare (channel-related) diseases, sample collection and specific therapeutic approaches are rather difficult. However, as described above, single cell samples from affected patients can be used to obtain iPSCs through reprogramming. These have the potential to be differentiated into different cell types and thus provide insights into the specific disease mechanisms of rare diseases. By modeling rare diseases with stem cells, potential target structures can be identified, and new therapeutic approaches can be developed, and researchers can target ways to restore the normal phenotype. This may involve the development of new drugs, targeted gene therapy, or the use of stem cell transplants [57].

However, in order to cope with genetic heterogeneity in patient tissues, it is advantageous to generate several cell clones from patients to be able to analyze them in parallel. As a control, clones from a close healthy family member are well suited. Additionally, these disease mutations should be corrected in the patient clones, whereas the disease-causing mutation should be introduced into the healthy clones. Via this approach, full control is given.

2.4 Electrophysiological characterization and drug testing

Stem cell-derived neurons can be studied electrophysiologically using patch-clamp techniques or multi-electrode arrays. These combined methods allow the direct measurement of action potentials, currents, and, in neurons, synaptic transmission to characterize the cells and their response to different stimuli. Thus, electrophysiological recordings are a prime tool to investigate drug effects directly on the targeted ion channel.

In this technique, a microelectrode is docked to the cell membrane to create a tight electrical contact. By applying a voltage difference and applying suction, a seal can be achieved between the electrode and the cell membrane. By applying a vacuum, a type of "patch" is created that allows the flow of ions to be measured across individual ion channels. The patch-clamp technique enables the measurement of membrane currents with high resolution and has contributed significantly to the understanding of ion channel function and signal transmission in cells [23, 63].

In cells, the membrane tightly regulates the osmolarity and maintains significant ion gradients, such as sodium (Na^+), potassium (K^+), chloride (Cl^-), and calcium (Ca^{2+}).

The approximate values for ion concentrations across human cell membranes are as follows:

1. Sodium (Na^+): The extracellular concentration of sodium ions is typically around 135–145 mM, while the intracellular concentration is lower, around 10–15 mM [64].

2. Potassium (K^+): The intracellular concentration of potassium ions is generally higher, ranging from 120 to 150 mM, while the extracellular concentration is typically around 3–5 mM [64].

3. Chloride (Cl^-): The extracellular concentration of chloride ions is around 95–110 mM, while the intracellular concentration is approximately 5–15 mM [64].

4. Calcium (Ca^{2+}): The extracellular concentration of calcium ions is around 1–2 mM, while the intracellular concentration is typically maintained at much lower levels, around 0.1–0.2 μM [64].

Overall, electrophysiological techniques such as the patch clamp technique or TEVC (two-electrode voltage clamp) are essential for investigating the functions of ion channels and the transport of ions across the membrane [65].

Extracellular derivation using microelectrode arrays or field potential amplifiers is a noninvasive method for electrophysiological characterization of cells. In this technique, electrodes are placed outside the cell to measure electric field potential emitted by the cell. Extracellular derivation allows the measurement of action potentials and synaptic activities of multiple cells simultaneously, providing a broader view of the interplay between cells [23, 65].

2.5 Understanding tissue-specific manifestations

As described above, channelopathies can affect various tissues and organs, each with its own unique cellular composition and physiology. Stem cell-based models allow researchers to generate different cell types affected by channelopathies and study their specific contributions to disease pathology. For example, patient-specific iPSCs can be differentiated into cardiomyocytes to study cardiac channelopathies or into neurons to investigate neurological channelopathies. These tissue-specific models provide a platform to elucidate the mechanisms underlying tissue-specific manifestations of channelopathies and develop targeted interventions for specific affected tissues [66, 67].

2.6 Development of cell-based therapies

In addition to drug screening, stem cell-based models hold promise for the development of cell-based therapies for channelopathies. By leveraging the differentiation potential of iPSCs, researchers can generate healthy, functional cells to replace or repair the affected tissues. For example, in cardiac channelopathies, iPSC-derived cardiomyocytes can be used as a source for cell transplantation to restore normal cardiac function – at least in mice for now [68]. Furthermore, the integration of gene editing technologies, such as CRISPR/Cas9, with stem cell-based models opens up possibilities for correcting disease-causing mutations and generating patient-specific healthy cells for autologous transplantation [69–71].

Cell replacement therapies and regenerative medicine are other promising aspects of stem cell applications in channelopathies. The ability of stem cells to differentiate into different cell types leads to the possibility of generating functional replacement cells that can be used in damaged tissues or organs. In the field of neuronal channelopathies, for example, stem cells could be used to generate healthy neurons to replace damaged areas in the brain or spinal cord [72]. Similarly, stem cells can be used to differentiate into cardiac cells to repair damaged heart tissue in cardiac channelopathies [73]. This approach offers promising prospects for the development of regenerative therapies that could provide long-term improvement in disease symptoms.

3. Advances in stem cell-based therapies for channelopathies

Stem cell-based therapy has made significant progress in the treatment of channelopathies in recent years. These genetic diseases are caused by mutations in the ion channels that lead to dysfunctional ion regulation. Here, we will focus on three important aspects of stem cell-based therapy for channelopathies: Gene editing and correction of disease-causing mutations, cell transplantation approaches, and challenges and future directions.

A promising approach in stem cell-based therapy for channelopathies is gene editing and the correction of disease-causing mutations. Through the development of techniques such as CRISPR/Cas9, gene editing allows the targeted modification of the genome of stem cells to correct specific mutations. Such approaches have been successfully applied to neuronal channelopathies, such as spinal muscular atrophy (SMA). Several studies showed that by correcting the disease-causing mutation in stem cells, the defective *SMN1* gene could be restored, resulting in improved motor neuron function [74, 75].

Another promising approach in stem cell-based therapy for channelopathies is the use of cell transplantation approaches. Here, differentiated stem cells or their derived cell types are transplanted into the affected tissue or organ to replace the dysfunctional cells or improve their function. In cardiac channelopathies such as long QT syndrome, it has been shown that transplantation of cardiac cells derived from stem cells can lead to an improvement in the electrical properties of the heart and a reduction in arrhythmias [76, 77].

Another example is the transplantation of encapsulated iPSC-derived β-cells competent for insulin production and glucose-level correlated release into type I diabetic pancreas of patients. This can improve the diabetic phenotype including in patients with defective insulin secretion as a consequence of defective β-cell ion channels.

In addition, stem cell-based therapy also enables a better understanding of the disease mechanisms and pathophysiology of channelopathies. By generating patient-specific induced pluripotent stem cells (iPSCs), researchers can replicate the affected cell types in the laboratory and study the effects of the disease-causing mutations on ion channel function and cell physiology. This understanding is crucial for the development of new therapeutic approaches. A study by Yazawa et al. [78] demonstrated the use of iPSCs to investigate the pathophysiological mechanisms of Timothy syndrome, a cardiac channelopathy.

However, despite the promising progress, there are also challenges and future directions in stem cell-based therapy for channelopathies. A major challenge is to generate differentiated stem cells in sufficient quantity and quality to use them for transplantation approaches. In addition, the safety and long-term efficacy of the transplanted cells must be ensured. Another future direction is to improve gene editing techniques to increase the efficiency and precision of mutation correction.

Overall, advances in stem cell-based therapy for channelopathies show promise for the treatment of these genetic diseases. Gene editing and correction of disease-causing mutations, as well as cell transplantation approaches, have the potential to alleviate the symptoms of channelopathies and improve the quality of life of affected patients. However, further research efforts and clinical trials are needed to confirm the safety, efficacy, and long-term effectiveness of these approaches.

3.1 Examples of channelopathies studied in iPSC-derived cells

Several channelopathies have been studied with the help of iPSC-derived cells. Some examples of prominent work will be briefly described hereafter.

3.1.1 Epilepsy

Research into epilepsies has benefited from the use of stem cell-derived neurons as a promising model. Stem cell-derived neurons offer the possibility to produce human neurons in the laboratory and thus to study epileptic phenomena in a controlled and reproducible system.

Stem cell-derived neurons can be used to recreate epileptic networks in the laboratory and explore the underlying mechanisms. For example, the study from Hirugashi et al. [79], in which stem cell-derived neurons were generated from patient cell samples with Dravet syndrome, a form of genetic epilepsy. The authors were able to show that these neurons were hyperexcitable and exhibited spontaneous epileptiform activity, similar to that seen in the brains of affected patients.

By using stem cell-derived neurons, researchers can more accurately investigate the specific disease mechanisms of epileptic disorders. Chen et al. [80] used stem cell-derived neurons from a cell line with a generated point mutation in *SCN1A*. This mutation leads to a misfunction in the $Na_V1.1$ α subunit, affecting neuronal function. The results provided important insights into the effects of these mutations on synaptic transmission and neuronal excitability.

KCNQ channels, also known as M channels, play a significant role in the regulation of neuronal excitability and have particular relevance in relation to epilepsy. KCNQ channels are a subset of voltage-gated potassium channels expressed in several regions of the brain, including the hippocampus, amygdala, and cortex. They are crucial for the regulation of neuronal excitability and contribute to the stabilization of resting membrane potential [18]. Mutations in the genes encoding KCNQ channels are associated with various forms of familial and sporadic epilepsy [81].

Studies have shown that mutations in the KCNQ channel genes can lead to impaired channel function, resulting in increased neuronal excitability and increased susceptibility to epileptic seizures. A particular form of epilepsy called benign familial neonatal epilepsy (BFNE) has been associated with mutations in the *KCNQ2* or *KCNQ3* gene [82].

The importance of KCNQ channels in the pathophysiology of epilepsy has sparked interest in developing therapies aimed at modulating these channels. One promising strategy is to identify pharmacological compounds that can enhance the activity of the channels and thus reduce neuronal hyperactivity. For example, a study by Wuttke et al. [83] investigated the effect of the compound retigabine on KCNQ channels and showed that it increased channel activity, thereby reducing neuronal hyperexcitability.

Also here, stem cells have helped to deepen the understanding of the role of KCNQ channels in the development of epilepsies. By differentiating stem cells into mature neurons – either directly or via neuronal progenitor cells – researchers can study the function and expression of KCNQ channels in a neuronal context. iPSC-derived neurons from patients with *KCNQ2*-associated epileptic encephalopathies have been used to investigate the effects of *KCNQ2* mutations on neuronal excitability [84, 85]. The results showed increased neuronal excitability and provided insights into the underlying mechanisms of the disease.

Overall, studies on KCNQ channels provide deeper insight into the pathophysiology of epilepsy and result in new possibilities for the development of therapeutic approaches. By using stem cells and studying patient-derived neurons, researchers can better understand the effects of KCNQ channel mutations and develop targeted therapies to control neuronal hyperactivity in epilepsy.

When it comes to pharmacology-based therapeutical approaches, studies have shown that the use of stem cells in the investigation of neuronal channelopathies such as epilepsy can provide crucial insights into the effects of antiepileptic drugs [85–87]. This could improve the effectiveness and safety of drugs and take into account individual differences in disease response [88, 89].

3.1.2 Long QT syndrome

Long QT syndrome (LQTS) is a hereditary condition characterized by a prolonged QT interval on the electrocardiogram (ECG) that can lead to life-threatening ventricular arrhythmias. One of the main causes of LQTS is channel dysfunction, particularly of ion channels responsible for regulating cardiac repolarization.

Several ion channels are associated with LQTS. One of them is the hERG (human ether-a-go-go related gene) channel. Mutations in this potassium channel lead to impaired repolarization of the action potential in the heart and a prolonged duration of the ventricular repolarization phase, which cause LQTS2 that increases the risk of torsade de pointes tachycardia. The function of the hERG channel is to allow the rapid outflow of potassium ions during the repolarization phase of the cardiac action potential [90].

A study examined the effects of three different mutations in the hERG gene on channel function [91]. Patch-clamp techniques were used to measure potassium current in stem cell-derived cardiac cells from patients with LQTS2 and healthy controls. They found that the LQTS2 patient mutations led to reduced hERG channel activity, resulting in prolonged repolarization time and increased risk of arrhythmias. These results support the role of hERG channel dysfunction in the pathogenesis of LQTS2 [90].

Additionally, *KCNQ1*, a gene encoding the KCNQ1 ion channel, can lead to impaired function of the channel and is associated with long QT syndrome 1, when mutated [92]. The KCNQ1 channel is responsible for regulating potassium ion flow during the repolarization phase of the action potential in the heart [93–96].

Another ion channel associated with LQTS is the *SCN5A* sodium channel. Mutations in this channel can lead to reduced sodium influx during the depolarization phase and impair repolarization of the action potential. This also leads to a prolonged QT interval (LQTS3) and an increased risk of arrhythmias [97]. Recent studies have investigated the function of *SCN5A* mutations in stem cell-derived cardiac cells and found that these mutations resulted in decreased sodium channel activity, which explains the impaired repolarization and prolonged QT time [98, 99].

iPS-derived cardiomyocytes have been used to study the electrophysiological behavior of single LQTS phenotypic cells [100–102]. For this purpose, the stem cells were first mutated to known LQTS genotypes, and then, the cells were differentiated and examined. In doing so, they examined not only LQTS but also other forms of arrhythmia. These cells could then be used in further steps to measure potential pharmaceuticals directly on the physiological cell systems and their immediate effect.

In summary, the hERG and KCNQ1 potassium channels as well as the SCN5A sodium channel play a central role in inherited LQTS 1-3. These channels are critical for the normal repolarization of the heart, and mutations in their genes lead to impaired ionic currents and prolonged repolarization, increasing the risk of life-threatening arrhythmias, and have been modeled in iPSC-derived cardiomyocytes.

Studies in stem cell-derived cardiac cells allow us to have a closer look at channel dysfunction and can be used as a screening platform to test direct therapeutic options at the physiological cellular level.

3.1.3 Periodic paralysis

Periodic paralysis, also known as paroxysmal paralysis, is a rare neurological disorder characterized by episodic weakness or paralysis of the muscles [13, 103]. These syndromes pose a challenge to the medical community because they are often difficult to diagnose and treat. Genetic mutations, particularly in the *CACNA1S*, *Kir2.1,* or *SCN4A* genes, have been linked to hypokalemic periodic paralysis and Andersen-Tawil syndrome [21, 103, 104].

The impaired function of ion channels in muscles leads to altered electrical excitability and impaired contractility of muscle fibers [19]. The underlying mechanisms of periodic paralysis are complex and can range from channel hyperpolarization to excessive depolarization [19]. This leads to impaired action potential formation and conduction, which in turn leads to characteristic paralysis attacks.

The use of stem cells, particularly induced pluripotent stem cells (iPSCs), has ushered in a new era also in the study of periodic paralysis [105, 106]. iPSCs can be produced from patient cells and allow the generation of specific cell types affected by the disease, such as muscle cells or neurons. These *in vitro* models allow researchers to study the effects of genetic mutations on the function of ion channels and the physiological properties of the affected cells.

The use of stem cells has not only contributed to a better understanding of the pathophysiology of periodic paralysis but also opened up new approaches for therapy development [106].

The study of periodic paralysis has already made and will make further significant progress thanks to advances in stem cell research and molecular genetic techniques. Further investigation of these disorders, particularly regarding their genetic diversity and the development of personalized therapies, will help to improve the quality of life of those affected.

4. Conclusion and future directions

Stem cell research has provided valuable insights into the understanding of channelopathies and offers promising avenues for the development of novel therapies. By generating disease-specific stem cell models, researchers have been able to study the underlying mechanisms of channelopathies and screen potential therapeutic agents. Additionally, stem cell-based approaches, such as gene editing and cell transplantation, hold great potential for the treatment of channelopathies. However, ethical considerations and regulatory frameworks need to be carefully addressed to ensure the responsible and ethical use of stem cells in research and clinical applications.

By specifically correcting genetic mutations in iPSCs or modulating ion channel activity in affected cells, potential treatment strategies can be explored. In addition, transplantation of healthy stem cells or differentiated cells into animal models provides new opportunities to study the efficacy of therapies.

Conflict of interest

The authors declare no conflict of interest.

Author details

Paul Disse*, Nadine Ritter, Nathalie Strutz-Seebohm and Guiscard Seebohm
Department of Cellular Electrophysiology, University Hospital Muenster, University of Muenster, Germany

*Address all correspondence to: paul.disse@uni-muenster.de

References

[1] Bernard G, Shevell MI. Channelopathies: A review. Pediatric Neurology. 2008;**38**:73-85. DOI: 10.1016/j.pediatrneurol.2007.09.007

[2] Catterall WA. Sodium channels, inherited epilepsy, and antiepileptic drugs. Annual Review of Pharmacology and Toxicology. 2014;**54**:317-338. DOI: 10.1146/annurev-pharmtox-011112-140232

[3] Fisher RS, Acevedo C, Arzimanoglou A, Bogacz A, Cross JH, Elger CE, et al. ILAE official report: A practical clinical definition of epilepsy. Epilepsia. 2014;**55**:475-482. DOI: 10.1111/epi.12550

[4] Nappi P, Miceli F, Soldovieri MV, Ambrosino P, Barrese V, Taglialatela M. Epileptic channelopathies caused by neuronal Kv7 (KCNQ) channel dysfunction. Pflugers Archiv European Journal of Physiology. 2020;**472**:881-898. DOI: 10.1007/s00424-020-02404-2

[5] Zerr P, Adelman JP, Maylie J. Episodic ataxia mutations in Kv1.1 alter potassium channel function by dominant negative effects or haploinsufficiency. The Journal of Neuroscience the Official Journal of the Society for Neuroscience. 1998;**18**:2842-2848. DOI: 10.1523/JNEUROSCI.18-08-02842.1998

[6] Fontaine B. Muscle channelopathies and related diseases. Handbook of Clinical Neurology. 2013;**113**:1433-1436. DOI: 10.1016/B978-0-444-59565-2.00012-5

[7] Nicole S, Fontaine B. Skeletal muscle sodium channelopathies. Current Opinion in Neurology. 2015;**28**:508-514. DOI: 10.1097/WCO.0000000000000238

[8] Kaufman ES. Mechanisms and clinical management of inherited channelopathies: Long QT syndrome, Brugada syndrome, catecholaminergic polymorphic ventricular tachycardia, and short QT syndrome. Heart Rhythm. 2009;**6**:S51-S55. DOI: 10.1016/j.hrthm.2009.02.009

[9] Corrado D, Basso C, Judge DP. Arrhythmogenic cardiomyopathy. Circulation Research. 2017;**121**:784-802. DOI: 10.1161/CIRCRESAHA.117.309345

[10] Zareba W, Cygankiewicz I. Long QT syndrome and short QT syndrome. Progress in Cardiovascular Diseases. 2008;**51**:264-278. DOI: 10.1016/j.pcad.2008.10.006

[11] Priori SG, Wilde AA, Horie M, Cho Y, Behr ER, Berul C, et al. HRS/EHRA/APHRS expert consensus statement on the diagnosis and management of patients with inherited primary arrhythmia syndromes: Document endorsed by HRS, EHRA, and APHRS in May 2013 and by ACCF, AHA, PACES, and AEPC in June 2013. Heart Rhythm. 2013;**10**:1932-1963. DOI: 10.1016/j.hrthm.2013.05.014

[12] Gnecchi M, Sala L, Schwartz PJ. Precision medicine and cardiac channelopathies: When dreams meet reality. European Heart Journal. 2021;**42**:1661-1675. DOI: 10.1093/eurheartj/ehab007

[13] Statland JM, Fontaine B, Hanna MG, Johnson NE, Kissel JT, Sansone VA, et al. Review of the diagnosis and treatment of periodic paralysis. Muscle & Nerves. 2018;**57**:522-530. DOI: 10.1002/mus.26009

[14] Gutmann L, Phillips LH. Myotonia congenita. Seminars

in Neurology. 1991;**11**:244-248. DOI: 10.1055/s-2008-1041228

[15] George AL. Inherited disorders of voltage-gated sodium channels. The Journal of Clinical Investigation. 2005;**115**:1990-1999. DOI: 10.1172/JCI25505

[16] Ptácek LJ. Channelopathies: Ion channel disorders of muscle as a paradigm for paroxysmal disorders of the nervous system. Neuromuscular Disorders. 1997;7:250-255. DOI: 10.1016/s0960-8966(97)000046-1

[17] Ashcroft FM. From molecule to malady. Nature. 2006;**440**:440-447. DOI: 10.1038/nature04707

[18] Jentsch TJ. Neuronal KCNQ potassium channels: Physiology and role in disease. Nature Reviews: Neuroscience. 2000;**1**:21-30. DOI: 10.1038/35036198

[19] Cannon SC. Channelopathies of skeletal muscle excitability. Comprehensive Physiology. 2015;**5**:761-790. DOI: 10.1002/cphy.c140062

[20] Zaydman MA, Silva JR, Cui J. Ion channel associated diseases: Overview of molecular mechanisms. Chemical Reviews. 2012;**112**:6319-6333. DOI: 10.1021/cr300360k

[21] Gillis J, Burashnikov E, Antzelevitch C, Blaser S, Gross G, Turner L, et al. Long QT, syndactyly, joint contractures, stroke and novel CACNA1C mutation: Expanding the spectrum of Timothy syndrome. American Journal of Medical Genetics. Part A. 2012;**158A**:182-187. DOI: 10.1002/ajmg.a.34355

[22] Tristani-Firouzi M, Etheridge SP. Electrical Diseases of the Heart. 1st ed. London: Springer; 2008. pp. 561-567. DOI: 10.1007/978-1-84628-854-8_37

[23] Hille B. Ion Channels of Excitable Membranes. 3rd ed. Sunderland: Sinauer Associates Inc; 2001

[24] Ruan Y, Denegri M, Liu N, Bachetti T, Seregni M, Morotti S, et al. Trafficking defects and gating abnormalities of a novel SCN5A mutation question gene-specific therapy in long QT syndrome type 3. Circulation Research. 2010;**106**:1374-1383. DOI: 10.1161/CIRCRESAHA.110.218891

[25] Misceo D, Holmgren A, Louch WE, Holme PA, Mizobuchi M, Morales RJ, et al. A dominant STIM1 mutation causes Stormorken syndrome. Human Mutation. 2014;**35**:556-564. DOI: 10.1002/humu.22544

[26] Catterall WA. From ionic currents to molecular mechanisms: The structure and function of voltage-gated sodium channels. Neuron. 2000;**26**:13-25. DOI: 10.1016/s0896-6273(00)81133-2

[27] Wei AD, Gutman GA, Aldrich R, Chandy KG, Grissmer S, Wulff H, et al. Nomenclature and molecular relationships of calcium-activated potassium channels. Pharmacological Reviews. 2005;**57**:463-472. DOI: 10.1124/pr.57.4.9

[28] Catterall WA. Signaling complexes of voltage-gated sodium and calcium channels. Neuroscience Letters. 2010;**486**:107-116. DOI: 10.1016/j.neulet.2010.08.085

[29] Abbott GW, Butler MH, Bendahhou S, Dalakas MC, Ptacek LJ, Goldstein SA. MiRP2 forms potassium channels in skeletal muscle with Kv3.4 and is associated with periodic paralysis. Cell. 2001;**104**:217-231. DOI: 10.1016/s0092-8674(01)00207-0

[30] Thomson JA, Itskovitz-Eldor J, Shapiro SS, Waknitz MA, Swiergiel JJ,

Marshall VS, et al. Embryonic stem cell lines derived from human blastocysts. Science. 1998;**282**:1145-1147. DOI: 10.1126/science.282.5391.1145

[31] Trounson A, McDonald C. Stem cell therapies in clinical trials: Progress and challenges. Cell Stem Cell. 2015;**17**:11-22. DOI: 10.1016/j.stem.2015.06.007

[32] Sanchez-Ramos J et al. Adult bone marrow stromal cells differentiate into neural cells in vitro. Experimental Neurology. 2000;**2000**:247-256. DOI: 10.1006/exnr.2000.7389

[33] Takahashi K, Yamanaka S. Induction of pluripotent stem cells from mouse embryonic and adult fibroblast cultures by defined factors. Cell. 2006;**126**:663-676. DOI: 10.1016/j.cell.2006.07.024

[34] Yu J, Vodyanik MA, Smuga-Otto K, Antosiewicz-Bourget J, Frane JL, Tian S, et al. Induced pluripotent stem cell lines derived from human somatic cells. Science. 2007;**318**:1917-1920. DOI: 10.1126/science.1151526

[35] Takahashi K, Yamanaka S. Induced pluripotent stem cells in medicine and biology. Development (Cambridge, England). 2013;**140**:2457-2461. DOI: 10.1242/dev.092551

[36] Karagiannis P, Takahashi K, Saito M, Yoshida Y, Okita K, Watanabe A, et al. Induced pluripotent stem cells and their use in human models of disease and development. Physiological Reviews. 2019;**99**:79-114. DOI: 10.1152/physrev.00039.2017

[37] Takahashi K, Yamanaka S. A decade of transcription factor-mediated reprogramming to pluripotency. Nature Reviews: Molecular Cell Biology. 2016;**17**:183-193. DOI: 10.1038/nrm.2016.8

[38] Kim K-P, Yoon J, Kim J, Röpke A, Shin B, Wook Han D, et al. Generation of a human iPSC line (MPIi007-A) from a patient with metachromatic leukodystrophy. Stem Cell Research. 2020;**48**:101993. DOI: 10.1016/j.scr.2020.101993

[39] Kim JB, Greber B, Araúzo-Bravo MJ, Meyer J, Park K, Zaehres H, et al. Direct reprogramming of human neural stem cells by OCT4. Nature. 2009;**461**:649-643. DOI: 10.1038/nature08436

[40] Hamill OP, Marty A, Neher E, Sakmann B, Sigworth FJ. Improved patch-clamp techniques for high-resolution current recording from cells and cell-free membrane patches. Pflugers Archiv European Journal of Physiology. 1981;**391**:85-100. DOI: 10.1007/BF00656997

[41] Noma A. ATP-regulated K+ channels in cardiac muscle. Nature. 1983;**305**:147-148. DOI: 10.1038/305147a0

[42] Bellin M, Marchetto MC, Gage FH, Mummery CL. Induced pluripotent stem cells: The new patient? Nature Reviews: Molecular Cell Biology. 2012;**13**:713-726. DOI: 10.1038/nrm3448

[43] Soldovieri MV, Miceli F, Taglialatela M. Driving with no brakes: molecular pathophysiology of Kv7 potassium channels. Physiology (Bethesda, Md.). 2011;**26**:365-376. DOI: 10.1152/physiol.00009.2011

[44] Renner H, Grabos M, Becker KJ, Kagermeier TE, Wu J, Otto M, et al. A fully automated high-throughput workflow for 3D-based chemical screening in human midbrain organoids. eLife. 2020;**9**:e52904. DOI: 10.7554/eLife.52904

[45] Liu Y, Lopez-Santiago LF, Yuan Y, Jones JM, Zhang H, O'Malley HA, et al.

Dravet syndrome patient-derived neurons suggest a novel epilepsy mechanism. Annals of Neurology. 2013;**74**:128-139. DOI: 10.1002/ana.23897

[46] Korn L, Speicher AM, Schroeter CB, Gola L, Kaehne T, Engler A, et al. MAPT genotype-dependent mitochondrial aberration and ROS production trigger dysfunction and death in cortical neurons of patients with hereditary FTLD. Redox Biology. 2023;**59**:102597. DOI: 10.1016/j.redox.2022.102597

[47] Voulgaris D, Nikolakopoulou P, Herland A. Generation of human iPSC-derived astrocytes with a mature star-shaped phenotype for CNS modeling. Stem Cell Reviews and Reports. 2022;**18**:2494-2512. DOI: 10.1007/s12015-022-10376-2

[48] Disse P, Aymanns I, Ritter N, Peischard S, Korn L, Wiendl H, et al. A novel NMDA receptor test model based on hiPSC-derived neural cells. Biological Chemistry. 2023;**404**:267-277. DOI: 10.1515/hsz-2022-0216

[49] Peischard S, Möller M, Disse P, Ho HT, Verkerk AO, Strutz-Seebohm N, et al. Virus-induced inhibition of cardiac pacemaker channel HCN4 triggers bradycardia in human-induced stem cell system. Cellular and Molecular Life Sciences CMLS. 2022;**79**:440. DOI: 10.1007/s00018-022-04435-7

[50] Schweizer PA, Darche FF, Ullrich ND, Geschwill P, Greber B, Rivinius R, et al. Subtype-specific differentiation of cardiac pacemaker cell clusters from human induced pluripotent stem cells. Stem Cell Research & Therapy. 2017;**8**:229. DOI: 10.1186/s13287-017-0681-4

[51] Zhao M-T, Shao N-Y, Garg V. Subtype-specific cardiomyocytes for precision medicine: Where are we now? Stem Cells (Dayton, Ohio). 2020;**38**:822-833. DOI: 10.1002/stem.3178

[52] Hendriks D, Clevers H, Artegiani B. CRISPR-Cas tools and their application in genetic engineering of human stem cells and organoids. Cell Stem Cell. 2020;**27**:705-731. DOI: 10.1016/j.stem.2020.10.014

[53] Moretti A, Bellin M, Welling A, Jung CB, Lam JT, Bott-Flügel L, et al. Patient-specific induced pluripotent stem-cell models for long-QT syndrome. The New England Journal of Medicine. 2010;**363**:1397-1409. DOI: 10.1056/NEJMoa0908679

[54] Manz KM, Siemann JK, McMahon DG, Grueter BA. Patch-clamp and multi-electrode array electrophysiological analysis in acute mouse brain slices. STAR Protocols. 2021;**2**:100442. DOI: 10.1016/j.xpro.2021.100442

[55] Hnatiuk AP, Briganti F, Staudt DW, Mercola M. Human iPSC modeling of heart disease for drug development. Cell Chemical Biology. 2021;**28**:271-282. DOI: 10.1016/j.chembiol.2021.02.016

[56] Soldner F, Hockemeyer D, Beard C, Gao Q, Bell GW, Cook EG, et al. Parkinson's disease patient-derived induced pluripotent stem cells free of viral reprogramming factors. Cell. 2009;**136**:964-977. DOI: 10.1016/j.cell.2009.02.013

[57] Dimos JT, Rodolfa KT, Niakan KK, Weisenthal LM, Mitsumoto H, Chung W, et al. Induced pluripotent stem cells generated from patients with ALS can be differentiated into motor neurons. Science. 2008;**321**:1218-1221. DOI: 10.1126/science.1158799

[58] Marchetto MCN, Carromeu C, Acab A, Yu D, Yeo GW, Mu Y, et al.

A model for neural development and treatment of Rett syndrome using human induced pluripotent stem cells. Cell. 2010;**143**:527-539. DOI: 10.1016/j.cell.2010.10.016

[59] Liu G-H, Suzuki K, Li M, Qu J, Montserrat N, Tarantino C, et al. Modelling Fanconi anemia pathogenesis and therapeutics using integration-free patient-derived iPSCs. Nature Communications. 2014;**5**:4330. DOI: 10.1038/ncomms5330

[60] Liang P, Lan F, Lee AS, Gong T, Sanchez-Freire V, Wang Y, et al. Drug screening using a library of human induced pluripotent stem cell-derived cardiomyocytes reveals disease-specific patterns of cardiotoxicity. Circulation. 2013;**127**:1677-1691. DOI: 10.1161/CIRCULATIONAHA.113.001883

[61] Freel BA, Sheets JN, Francis KR. iPSC modeling of rare pediatric disorders. Journal of Neuroscience Methods. 2020;**332**:108533. DOI: 10.1016/j.jneumeth.2019.108533

[62] Vatine GD, Barrile R, Workman MJ, Sances S, Barriga BK, Rahnama M, et al. Human iPSC-derived blood-brain barrier chips enable disease modeling and personalized medicine applications. Cell Stem Cell. 2019;**24**:995-1005.e6. DOI: 10.1016/j.stem.2019.05.011

[63] Neher E, Sakmann B. Single-channel currents recorded from membrane of denervated frog muscle fibres. Nature. 1976;**260**:799-802. DOI: 10.1038/260799a0

[64] Albers B, Johnson A, Lewis J, Raff M, Roberts K, Walter P. Molecular Biology of the Cell. 6th ed. New York: Garland Science; 2015

[65] Bell DC, Dallas ML. Advancing Ion Channel research with automated patch clamp (APC) electrophysiology platforms. Advances in Experimental Medicine and Biology. 2021;**1349**:21-32. DOI: 10.1007/978-981-16-4254-8_2

[66] Bellin M, Casini S, Davis RP, D'Aniello C, Haas J, Ward-van Oostwaard D, et al. Isogenic human pluripotent stem cell pairs reveal the role of a KCNH2 mutation in long-QT syndrome. The EMBO Journal. 2013;**32**:3161-3175. DOI: 10.1038/emboj.2013.240

[67] Wainger BJ, Buttermore ED, Oliveira JT, Mellin C, Lee S, Saber WA, et al. Modeling pain in vitro using nociceptor neurons reprogrammed from fibroblasts. Nature Neuroscience. 2015;**18**:17-24. DOI: 10.1038/nn.3886

[68] Jiang X, Yang Z, Dong M. Cardiac repair in a murine model of myocardial infarction with human induced pluripotent stem cell-derived cardiomyocytes. Stem Cell Research & Therapy. 2020;**11**:297. DOI: 10.1186/s13287-020-01811-7

[69] Liang P, Xu Y, Zhang X, Ding C, Huang R, Zhang Z, et al. CRISPR/Cas9-mediated gene editing in human tripronuclear zygotes. Protein & Cell. 2015;**6**:363-372. DOI: 10.1007/s13238-015-0153-5

[70] Mandai M, Watanabe A, Kurimoto Y, Hirami Y, Morinaga C, Daimon T, et al. Autologous induced stem-cell-derived retinal cells for macular degeneration. The New England Journal of Medicine. 2017;**376**:1038-1046. DOI: 10.1056/NEJMoa1608368

[71] Li C, Chen S, Zhou Y, Zhao Y, Liu P, Cai J. Application of induced pluripotent stem cell transplants: Autologous or allogeneic? Life Sciences. 2018;**212**:145-149. DOI: 10.1016/j.lfs.2018.09.057

[72] Vierbuchen T, Ostermeier A, Pang ZP, Kokubu Y, Südhof TC, Wernig M. Direct conversion of fibroblasts to functional neurons by defined factors. Nature. 2010;**463**:1035-1041. DOI: 10.1038/nature08797

[73] Lahti AL, Kujala VJ, Chapman H, Koivisto A-P, Pekkanen-Mattila M, Kerkelä E, et al. Model for long QT syndrome type 2 using human iPS cells demonstrates arrhythmogenic characteristics in cell culture. Disease Models & Mechanisms. 2012;**5**:220-230. DOI: 10.1242/dmm.008409

[74] Feng M, Liu C, Xia Y, Liu B, Zhou M, Li Z, et al. Restoration of SMN expression in mesenchymal stem cells derived from gene-targeted patient-specific iPSCs. Journal of Molecular Histology. 2018;**49**:27-37. DOI: 10.1007/s10735-017-9744-1

[75] Zhou M, Tang S, Duan N, Xie M, Li Z, Feng M, et al. Targeted-deletion of a tiny sequence via prime editing to restore SMN expression. International Journal of Molecular Sciences. 2022:23. DOI: 10.3390/ijms23147941

[76] Guo R, Morimatsu M, Feng T, Lan F, Chang D, Wan F, et al. Stem cell-derived cell sheet transplantation for heart tissue repair in myocardial infarction. Stem Cell Research & Therapy. 2020;**11**:19. DOI: 10.1186/s13287-019-1536-y

[77] Schwartz PJ, Gnecchi M, Dagradi F, Castelletti S, Parati G, Spazzolini C, et al. From patient-specific induced pluripotent stem cells to clinical translation in long QT syndrome Type 2. European Heart Journal. 2019;**40**:1832-1836. DOI: 10.1093/eurheartj/ehz023

[78] Yazawa M, Hsueh B, Jia X, Pasca AM, Bernstein JA, Hallmayer J, et al. Using induced pluripotent stem cells to investigate cardiac phenotypes in Timothy syndrome. Nature. 2011;**471**:230-234. DOI: 10.1038/nature09855

[79] Higurashi N, Uchida T, Lossin C, Misumi Y, Okada Y, Akamatsu W, et al. A human Dravet syndrome model from patient induced pluripotent stem cells. Molecular Brain. 2013;**6**:19. DOI: 10.1186/1756-6606-6-19

[80] Chen W, Liu J, Zhang L, Xu H, Guo X, Deng S, et al. Generation of the SCN1A epilepsy mutation in hiPS cells using the TALEN technique. Scientific Reports. 2014;**4**:5404. DOI: 10.1038/srep05404

[81] Greene DL, Hoshi N. Modulation of Kv7 channels and excitability in the brain. Cellular and Molecular Life Sciences CMLS. 2017;**74**:495-508. DOI: 10.1007/s00018-016-2359-y

[82] Biervert C, Schroeder BC, Kubisch C, Berkovic SF, Propping P, Jentsch TJ, et al. A potassium channel mutation in neonatal human epilepsy. Science. 1998;**279**:403-406. DOI: 10.1126/science.279.5349.403

[83] Wuttke TV, Jurkat-Rott K, Paulus W, Garncarek M, Lehmann-Horn F, Lerche H. Peripheral nerve hyperexcitability due to dominant-negative KCNQ2 mutations. Neurology. 2007;**69**:2045-2053. DOI: 10.1212/01.wnl.0000275523.95103.36

[84] Simkin D, Marshall KA, Vanoye CG, Desai RR, Bustos BI, Piyevsky BN, et al. Dyshomeostatic modulation of Ca2+−activated K+ channels in a human neuronal model of KCNQ2 encephalopathy. eLife. 2021;**2021**:10. DOI: 10.7554/eLife.64434

[85] Simkin D, Ambrosi C, Marshall KA, Williams LA, Eisenberg J, Gharib M, et al. 'Channeling' therapeutic discovery

for epileptic encephalopathy through iPSC technologies. Trends in Pharmacological Sciences. 2022;**43**:392-405. DOI: 10.1016/j.tips.2022.03.001

[86] Mariani J, Simonini MV, Palejev D, Tomasini L, Coppola G, Szekely AM, et al. Modeling human cortical development in vitro using induced pluripotent stem cells. Proceedings of the National Academy of Sciences of the United States of America. 2012;**109**:12770-12775. DOI: 10.1073/pnas.1202944109

[87] Lybrand ZR, Goswami S, Hsieh J. Stem cells: A path towards improved epilepsy therapies. Neuropharmacology. 2020;**168**:107781. DOI: 10.1016/j.neuropharm.2019.107781

[88] Braam SR, Denning C, van den Brink S, Kats P, Hochstenbach R, Passier R, et al. Improved genetic manipulation of human embryonic stem cells. Nature Methods. 2008;**5**:389-392. DOI: 10.1038/nmeth.1200

[89] Hirose S, Tanaka Y, Shibata M, Kimura Y, Ishikawa M, Higurashi N, et al. Application of induced pluripotent stem cells in epilepsy. Molecular and Cellular Neurosciences. 2020;**108**:103535. DOI: 10.1016/j.mcn.2020.103535

[90] Mehta A, Sequiera GL, Ramachandra CJA, Sudibyo Y, Chung Y, Sheng J, et al. Re-trafficking of hERG reverses long QT syndrome 2 phenotype in human iPS-derived cardiomyocytes. Cardiovascular Research. 2014;**102**:497-506. DOI: 10.1093/cvr/cvu060

[91] Keller DI, Grenier J, Christé G, Dubouloz F, Osswald S, Brink M, et al. Characterization of novel KCNH2 mutations in type 2 long QT syndrome manifesting as seizures. The Canadian Journal of Cardiology. 2009;**25**:455-462. DOI: 10.1016/s0828-282x(09)70117-5

[92] Matsuda S, Ohnuki Y, Okami M, Ochiai E, Yamada S, Takahashi K, et al. Jervell and Lange-Nielsen syndrome with novel KCNQ1 and additional gene mutations. Human Genome Variation. 2020;**7**:34. DOI: 10.1038/s41439-020-00121-x

[93] Barhanin J, Lesage F, Guillemare E, Fink M, Lazdunski M, Romey G. K(V) LQT1 and lsK (minK) proteins associate to form the I(Ks) cardiac potassium current. Nature. 1996;**384**:78-80. DOI: 10.1038/384078a0

[94] Moss AJ, Kass RS. Long QT syndrome: From channels to cardiac arrhythmias. The Journal of Clinical Investigation. 2005;**115**:2018-2024. DOI: 10.1172/JCI25537

[95] Bellin M, Greber B. Human iPS cell models of Jervell and Lange-Nielsen syndrome. Rare Diseases. 2015;**3**:e1012978. DOI: 10.1080/21675511.2015.1012978

[96] Zhang M, D'Aniello C, Verkerk AO, Wrobel E, Frank S, Ward-van Oostwaard D, et al. Recessive cardiac phenotypes in induced pluripotent stem cell models of Jervell and Lange-Nielsen syndrome: Disease mechanisms and pharmacological rescue. Proceedings of the National Academy of Sciences of the United States of America. 2014;**111**:E5383-E5392. DOI: 10.1073/pnas.1419553111

[97] Wilde AAM, Amin AS. Clinical spectrum of SCN5A mutations: Long QT syndrome, Brugada syndrome, and cardiomyopathy. JACC. Clinical Electrophysiology. 2018;**4**:569-579. DOI: 10.1016/j.jacep.2018.03.006

[98] Li W, Stauske M, Luo X, Wagner S, Vollrath M, Mehnert CS, et al. Disease phenotypes and mechanisms of iPSC-derived Cardiomyocytes from

Brugada syndrome patients with a loss-of-function SCN5A mutation. Frontiers in Cell and Developmental Biology. 2020;**8**:592893. DOI: 10.3389/fcell.2020.592893

[99] La Roche J, Angsutararux P, Kempf H, Janan M, Bolesani E, Thiemann S, et al. Comparing human iPSC-cardiomyocytes versus HEK293T cells unveils disease-causing effects of Brugada mutation A735V of NaV1.5 sodium channels. Scientific Reports. 2019;**9**:11173. DOI: 10.1038/s41598-019-47632-4

[100] Sinnecker D, Goedel A, Dorn T, Dirschinger RJ, Moretti A, Laugwitz K-L. Modeling long-QT syndromes with iPS cells. Journal of Cardiovascular Translational Research. 2013;**6**:31-36. DOI: 10.1007/s12265-012-9416-1

[101] Sinnecker D, Goedel A, Laugwitz K-L, Moretti A. Induced pluripotent stem cell-derived cardiomyocytes: A versatile tool for arrhythmia research. Circulation Research. 2013;**112**:961-968. DOI: 10.1161/CIRCRESAHA.112.268623

[102] Sinnecker D, Laugwitz K-L, Moretti A. Induced pluripotent stem cell-derived cardiomyocytes for drug development and toxicity testing. Pharmacology & Therapeutics. 2014;**143**:246-252. DOI: 10.1016/j.pharmthera.2014.03.004

[103] Matthews E, Hanna MG. Muscle channelopathies: Does the predicted channel gating pore offer new treatment insights for hypokalaemic periodic paralysis? The Journal of Physiology. 2010;**588**:1879-1886. DOI: 10.1113/jphysiol.2009.186627

[104] Wu F, Mi W, Fu Y, Cannon SC. Mice with a null allele for NaV1.4 exhibit Pseudo-myasthenia, but are not susceptibile to periodic paralysis. Biophysical Journal. 2016;**110**:111a. DOI: 10.1016/j.bpj.2015.11.656

[105] Dee CT, Szymoniuk CR, Mills PED, Takahashi T. Defective neural crest migration revealed by a zebrafish model of Alx1-related frontonasal dysplasia. Human Molecular Genetics. 2013;**22**:239-251. DOI: 10.1093/hmg/dds423

[106] Pini J, Rouleau M, Desnuelle C, Sacconi S, Bendahhou S. Modeling Andersen's syndrome in human induced pluripotent stem cells. Stem Cells and Development. 2016;**25**:151-159. DOI: 10.1089/scd.2015.0258